Electrical installation calculations

Volume 2

Electrical installation calculations

VOLUME 2

SIXTH EDITION

A. J. Watkins

Chris Kitcher

AMSTERDAM • BOSTON • HEIDELBERG • LONDON • NEW YORK
OXFORD • PARIS • SAN DIEGO • SAN FRANCISCO
SINGAPORE • SYDNEY • TOKYO
Newnes is an imprint of Elsevier
ELSEVIER

Newnes

Newnes is an imprint of Elsevier
Linacre House, Jordan Hill, Oxford OX2 8DP
30 Corporate Drive, Suite 400, Burlington, MA 01803

First edition 1957
Fifth edition 1999
Reprinted 2001, 2002, 2003, 2004
Sixth edition 2006

British Library Cataloguing in Publication Data
A catalogue record for this book is available from the British Library

Library of Congress Cataloging-in-Publication Data
A catalog record for this book is available from the Library of Congress

ISBN–13 978-0-7506-6783-8
ISBN–10 0-7506-6783-4

For information on all Newnes publications
visit our website at www.newnespress.com

Printed and bound in UK
06 07 08 09 10   10 9 8 7 6 5 4 3 2 1

Working together to grow
libraries in developing countries

www.elsevier.com | www.bookaid.org | www.sabre.org

ELSEVIER   BOOK AID
International   Sabre Foundation

Contents

Preface

Mathematics forms the essential foundation of electrical installation work. Without applying mathematical functions we would be unable to work out the size of a room which needs lighting or heating, the size and/or the number of the lights or heaters themselves, the number and/or the strength of the fixings required, or the size of the cables supplying them. We would be unable to accurately establish the rating of the fuse or circuit breaker needed to protect the circuits, or predict the necessary test results when testing the installation. Like it or not you will need to be able to carry out mathematics if you want to be an efficient and skilled electrician.

This book will show you how to perform the maths you will need to be a proficient electrician. It concentrates on the electronic calculator methods you would use in class and in the workplace. The book does not require you to have a deep understanding of how the mathematical calculations are performed; you are taken through each topic step by step, then you are given the opportunity yourself to carry out exercises at the end of each chapter. Throughout the book useful references are made to Amendment 2:2004 BS 7671 Requirements for Electrical Regulations and the *IEE On-Site Guide*.

Volume 2 Electrical Installation Calculations originally written by A. J. Watkins and R. K. Parton has been the preferred book for many students looking to improve their mathematical understanding of the subject for many years. This edition has been newly updated not only to include modern methods, but also to cover all aspects of the new City and Guilds 2330 Certificate in Electrotechnical Technology.

This second volume includes advanced calculations, in particular those involving cable selection. As well as being invaluable to students studying for the City and Guilds 2330, it will also be of considerable use to those involved in electrical installation work, particularly if studying for the City and Guilds 2391 Inspection and Testing, 2400 Design and Verification and the 2381 exams.

Chris Kitcher

Use of calculators

Throughout Books 1 and 2, the use of a calculator is encouraged. Your calculator is a tool, and like any tool practice is required to perfect its use. A scientific calculator will be required, and although they differ in the way the functions are carried out, the end result is the same.

The examples are given using a Casio fx-83MS. The figures printed on the button is the function performed when the button is pressed. To use the function in small letters above any button the *shift* button must be used.

PRACTICE IS IMPORTANT

Syntax error	Appears when the figures are entered in the wrong order.
x^2	Multiplies a number by itself, i.e. $6 \times 6 = 36$. On the calculator this would be $6x^2 = 36$. When a number is multiplied by itself it is said to be *squared*.
x^3	Multiplies a number by itself and then the total by itself again, i.e. when we enter 4 on calculator $x^3 = 64$. When a number is multiplied in this way it is said to be *cubed*.
$\sqrt{}$	Gives the number which achieves the total by being multiplied by itself, i.e. $\sqrt{36} = 6$. This is said to be the *square root* of a number and is the opposite of *squared*.
$\sqrt[3]{}$	Gives you the number which when multiplied by itself three times will be the total. $\sqrt[3]{64} = 4$ this is said to be the *cube root*.

x^{-1} Divides 1 by a number, i.e. $\frac{1}{4} = 0.25$ This is the *reciprocal* button and is useful in this book for finding the resistance of resistors in parallel and capacitors in series.

EXP   The powers of 10 function, i.e.

$25 \times 1000 = 25\ \text{EXP} \times 10^3 = 25\,000$

Enter into calculator 25 EXP 3 = 25 000. (Do not enter the × or the number 10.)

If a calculation shows 10^{-3}, i.e. 25×10^{-3} enter 25 EXP -3 = (0.025) (when using EXP if a minus is required use the button $(-)$)

Brackets   These should be used to carry out a calculation within a calculation. Example calculation:

$$\frac{32}{(0.8 \times 0.65 \times 0.94)} = 65.46$$

Enter into calculator $32 \div (0.8 \times 0.65 \times 0.94) =$

Remember, *Practice makes perfect!*

Simple transposition of formulae

To find an unknown value:

- The subject must be on the top line and must be on its own.
- The answer will always be on the top line.
- To get the subject on its own, values must be moved.
- Any value that moves across the = sign must move

  from above the line to below line or

  from below the line to above the line.

$$\frac{3 \times 4 = 2 \times 6}{3 \times 4 = 2 \times ?}$$

Transpose to find ?

$$\frac{3 \times 4}{2} = 6$$

$$\frac{2 \times 6}{?} = 4$$

Step 1 $\quad \dfrac{2 \times 6 = 4 \times ?}{}$

Step 2 $\quad \dfrac{2 \times 6}{4} = ?$

Answer $\quad \dfrac{2 \times 6}{4} = 3$

EXAMPLE 3

$$5 \times 8 \times 6 = 3 \times 20 \times \; ?$$

Step 1: move 3×20 away from the unknown value, as the known values move across the $=$ sign they must move to the bottom of the equation

$$\frac{5 \times 8 \times 4}{3 \times 20} = \; ?$$

Step 2: Carry out the calculation

$$\frac{5 \times 8 \times 6}{3 \times 20} = \frac{240}{60} = 4$$

Therefore

$$5 \times 8 \times 6 = 240$$

$$3 \times 20 \times 4 = 240$$

or

$$5 \times 8 \times 6 = 3 \times 20 \times 4.$$

SI units

In Europe and the UK, the units for measuring different properties are known as SI units.

SI stands for *Système Internationale.*

All units are derived from seven base units.

Base quantity	Base unit	Symbol
Time	Second	s
Electrical current	Ampere	A
Length	Metre	m
Mass	Kilogram	kg
Temperature	Kelvin	K
Luminous intensity	Candela	cd
Amount of substance	Mole	mol

SI-DERIVED UNITS

Derived quantity	Name	Symbol
Frequency	Hertz	Hz
Force	Newton	N
Energy, work, quantity of heat	Joule	J
Electric charge, quantity of electricity	Coulomb	C
Power	Watt	W
Potential difference, electromotive force	Volt	V or U
Capacitance	Farad	F
Electrical resistance	Ohm	Ω
Magnetic flux	Weber	Wb
Magnetic flux density	Tesla	T

(Continued)

Derived quantity	Name	Symbol
Inductance	Henry	H
Luminous flux	Lumen	cd
Area	Square metre	m^2
Volume	Cubic metre	m^3
Velocity, speed	Metre per second	m/s
Mass density	Kilogram per cubic metre	kg/m^3
Luminance	Candela per square metre	cd/m^2

SI UNIT PREFIXES

Name	Multiplier	Prefix	Power of 10
Tera	1000 000 000 000	T	1×10^{12}
Giga	1000 000 000	G	1×10^9
Mega	1000 000	M	1×10^6
Kilo	1000	k	1×10^3
Unit	1		
Milli	0.001	m	1×10^{-3}
Micro	0.000 001	μ	1×10^{-6}
Nano	0.000 000 001	η	1×10^{-9}
Pico	0.000 000 000 001	ρ	1×10^{-12}

EXAMPLE

mA	Milliamp = one thousandth of an ampere
km	Kilometre = one thousand metres
μv	Microvolt = one millionth of a volt
GW	Gigawatt = one thousand million watts
kW	Kilowatt = one thousand watts

Calculator example

1 kilometre is 1 metre $\times 10^3$

Enter into calculator 1 EXP 3 = (1000) metres

1000 metres is 1 kilometre $\times 10^{-3}$

Enter into calculator 1000 EXP −3 = (1) kilometre

1 microvolt is 1 volt $\times 10^{-6}$

Enter into calculator 1 EXP −6 = (1^{-06} or 0.000 001) volts
(note sixth decimal place).

Conductor colour identification

	Old colour	New colour	Marking
Phase 1 of a.c.	Red	Brown	L1
Phase 2 of a.c.	Yellow	Black	L2
Phase 3 of a.c.	Blue	Grey	L3
Neutral of a.c.	Black	Blue	N

Note: great care must be taken when working on installations containing old and new colours.

Alternating current circuit calculations

IMPEDANCE

In d.c. circuits, the current is limited by resistance. In a.c. circuits, the current is limited by *impedance* (Z). Resistance and impedance are measured in ohms.

For this calculation, Ohm's law is used and Z is substituted for R.

$$\frac{U}{Z} = I \text{ or voltage } (U) \div \text{ impedance (ohms)}$$

$$= \text{current (amperes)}$$

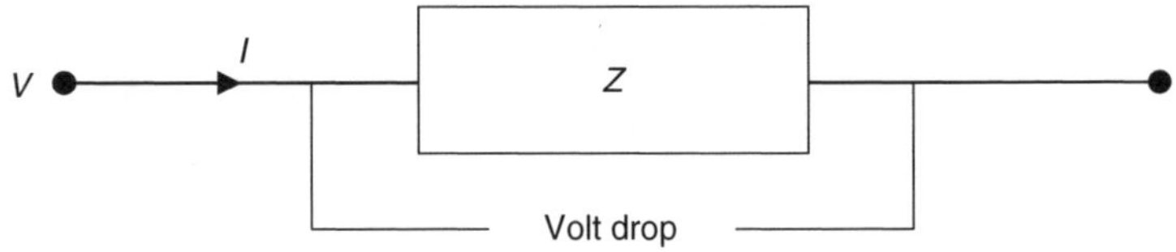

Fig. 1

EXAMPLE I The voltage applied to a circuit with an impedance of $6\,\Omega$ is 200 volts. Calculate the current in the circuit.

$$\frac{U}{Z} = I$$

$$\frac{200}{6} = 33.33 \text{ A}$$

EXAMPLE 2   The current in a 230 V single phase motor is 7.6 A. Calculate the impedance of the circuit.

$$\frac{U}{I} = Z$$

$$\frac{230}{7.6} = 30.26\,\Omega$$

EXAMPLE 3   A discharge lamp has an impedance of 265 Ω and the current drawn by the lamp is 0.4 A. Calculate the voltage.

$$Z \times I = U$$

$$265 \times 0.4 = 110 \text{ volts}$$

EXAMPLE 4   The current through an impedance of 32 Ω is 8 A. Calculate the voltage drop across the impedance.

$$U = I \times Z$$
$$= 8 \times 32$$
$$= 256 \text{ V}$$

EXAMPLE 5   The current through an electric motor is 6.8 A at 230 V. Calculate the impedance of the motor.

$$U = I \times Z$$
$$(\text{Transpose for } Z)\ Z = \frac{U}{I}$$
$$= \frac{230}{6.8}$$
$$= 33.82\,\Omega$$

EXAMPLE 6   An a.c. coil has an impedance of 430 Ω. Calculate the voltage if the coil draws a current of 0.93 A.

$$U = I \times Z$$
$$(\text{Transpose for } U)\ U = I \times Z$$
$$= 0.93 \times 430$$
$$= 400 \text{ V}$$

1. Complete the following table:

Volts (a.c.)		230	400		100	25		230	
Current (A)	0.1	15	0.5			0.01		180	25
Impedance (Ω)	100	15			1000		0.05		25

2. Complete the following table:

Current (A)	1.92	3.84	18.2		7.35	4.08		8.97
Volts (a.c.)		7.5		230	107		400	235
Impedance (Ω)	2.45		12.4	96.3		56	96	

3. Complete the following table:

Impedance (Ω)	232	850		0.125			1050	129
Volts (a.c.)		230	400	26.5	0.194		238	245
Current (A)	0.76		0.575		0.0065	0.436		0.056

4. A mercury vapour lamp takes 2.34 A when the mains voltage is 237 V. Calculate the impedance of the lamp circuit.

5. An inductor has an impedance of 365 Ω. How much current will flow when it is connected to a 400 V a.c. supply?

6. A coil of wire passes a current of 55 A when connected to a 120 V d.c. supply but only 24.5 A when connected to a 110 V a.c. supply. Calculate (a) the resistance of the coil, (b) its impedance.

7. Tests to measure the impedance of an earth fault loop were made in accordance with BS 7671 and the results for five different installations are given below. For each case, calculate the value of the loop impedance.

	Test voltage, a.c. (V)	Current (A)
(a)	9.25	19.6
(b)	12.6	3.29
(c)	7.65	23.8
(d)	14.2	1.09
(e)	8.72	21.1

8. The choke in a certain fluorescent-luminaire fitting causes a voltage drop of 150 V when the current through it is 1.78 A. Calculate the impedance of the choke.

9. Complete the following table:

Volts (a.c.)	61.1		153	193	
Current (A)	2.3	4.2		7.35	9.2
Impedance (Ω)		25	25		25

Plot a graph showing the relationship between current and voltage. From the graph, state the value of the current when the voltage is 240 V.

10. The alternating voltage applied to a circuit is 230 V and the current flowing is 0.125 A. The impedance of the circuit is
(a) 5.4 Ω (b) 1840 Ω (c) 3.5 Ω (d) 184 Ω

11. An alternating current of 2.4 A flowing in a circuit of impedance 0.18 Ω produces a voltage drop of
(a) 0.075 V   (b) 13.3 V   (c) 0.432 V   (d) 4.32 V

12. When an alternating e.m.f. of 150 V is applied to a circuit of impedance 265 Ω, the current is
(a) 39 750 A   (b) 1.77 A   (c) 5.66 A   (d) 0.566 A

INDUCTIVE REACTANCE

When an a.c. current is passed through a conductor, a magnetic field is created around the conductor. If the conductor is wound into a coil the magnetic field is increased. Where there are significant magnetic fields in a circuit there is opposition to the flow of current, this opposition is called *inductive reactance*. The opposition caused by inductive reactance is in addition to the opposition caused by the resistance caused by the wires.

In this section, we will assume that the resistance of the circuits is so low that it may be ignored and that the only opposition to the flow of current is that caused by the inductive reactance.

The formulae for inductive reactance

$$X_L = 2\pi fL \text{ (answer in ohms)}.$$

Where L is the inductance of the circuit or coil of wire and is stated in henrys (H), f is the frequency of the supply in hertz (Hz).

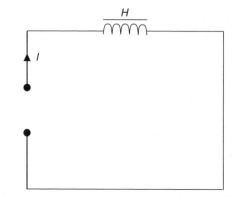

Fig. 2

EXAMPLE 1   Calculate the inductive reactance of a coil which has an inductance of 0.03 henrys when connected to a 50 Hz supply.

$$X_L = 2\pi f L$$
$$= 2 \times 3.142 \times 50 \times 0.03 = 9.42\ \Omega$$

EXAMPLE 2   Calculate the inductive reactance of the coil in example 1 when connected to a 60 Hz supply.

$$X_L = 2\pi f L$$
$$= 2 \times 3.142 \times 60 \times 0.03 = 11.31\ \Omega$$

It can be seen from this calculation that if the frequency increases the inductive reactance will also increase.

EXAMPLE 3   An inductor is required to cause a voltage drop of 180 volts when a current of 1.5 A is passed through it at a frequency of 50 Hz.

Calculate the value of the inductor:

$$U_L = I \times X_L \text{ (this is Ohm's law with inductive}$$
$$\text{reactance instead of resistance)}$$

Transposed

$$\frac{U}{I} = X_L$$

$$\frac{180}{1.5} = 120\,\Omega$$

$$X_L = 2\pi f L$$

$$120 = 2 \times 3.142 \times 50 \times L$$

Transpose

$$\frac{120}{(2 \times 3.142 \times 50)} = 0.381\,\text{H}$$

On calculator enter $120 \div (2\pi \times 50) =$ (answer 0.382 H)

EXERCISE 2

1. Calculate the inductive reactance of a coil having an inductance of 0.015 H when a 50 Hz current flows in it.
2. A coil is required to have an inductive reactance of 150 Ω on a 50 Hz supply. Determine its inductance.
3. Complete the following table:

Inductance (H)	0.04		0.12	0.008	
Frequency (Hz)	50	50			60
Reactance (Ω)		50	36	4.5	57

4. A coil of negligible resistance causes a voltage drop of 98 V when the current through it is 2.4 A at 50 Hz. Calculate (a) the reactance of the coil, (b) its inductance.
5. A reactor allows a current of 15 A to flow from a 230 V 50 Hz supply. Determine the current which will flow at the same voltage if the frequency changes to (a) 45 Hz, (b) 55 Hz. Ignore the resistance.
6. Calculate the inductive reactance of coils having the following values of inductance when the supply frequency is 50 Hz.
   (a) 0.012 H
   (b) 0.007 H
   (c) 0.45 mH
   (d) 350 μH
   (e) 0.045 H

7. Determine the inductances of the coils which will have the following reactance to a 50 Hz supply:
   (a) 300 Ω   (d) 125 Ω
   (b) 25 Ω   (e) 5 Ω
   (c) 14.5 Ω
8. The inductance of a coil can be varied from 0.15 H to 0.06 H. Plot a graph to show how the inductive reactance varies with changes in inductance. Assume a constant frequency of 50 Hz.
9. A reactor has a constant inductance of 0.5 H and it is connected to a supply of constant voltage 100 V but whose frequency varies from 25 to 50 Hz. Plot a graph to show how the current through the coil changes according to the frequency. Ignore the resistance of the coil.
10. Calculate the voltage drop across a 0.24 H inductor of negligible resistance when it carries 5.5 A at 48 Hz.
11. An inductor of 0.125 H is connected to an a.c. supply at 50 Hz. Its inductive reactance is
    (a) 39.3 Ω   (b) 0.79 Ω   (c) 0.025 Ω   (d) 393 Ω
12. The value in henrys of an inductor which has an inductive reactance of 500 Ω when connected in an a.c. circuit at frequency 450 Hz is
    (a) 1.77 H   (c) 0.177 H
    (b) 14×10^6 H   (d) 0.071×10^{-6} H

CAPACITIVE REACTANCE

When a capacitor is connected to an a.c. supply, the current flow is limited by the *reactance* of the capacitor (X_C).

$$\text{Formula for capacitive reactance } X_C = \frac{10^6}{2\pi fC}$$

where C is the capacitance of the capacitor measured in microfarads (μF) and f is the frequency of the supply in hertz (Hz). (It should be noted that d.c. current will not flow with a capacitor in the circuit it will simply charge and then stop.)

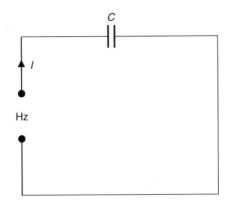

Fig. 3

EXAMPLE 1   Calculate the reactance of a $70\,\mu\text{F}$ capacitor to a 50 Hz supply:

$$X_C = \frac{10^6}{2\pi fC}$$

$$\frac{10^6}{2 \times 3.142 \times 50 \times 70} = 45.47\,\Omega$$

Enter on calculator EXP 6 $\div$ ($2\pi \times 50 \times 70$) = (answer 45.47).

EXAMPLE 2   A power factor improvement capacitor is required to take a current of 7.5 A from a 230 volt 50 Hz supply.

   Determine the value of the capacitor.

   For this calculation, Ohm's law is used and R is substituted by X_C.

Step 1

$$U_C = I \times X_C$$

$$230 = 7.5 \times X_C$$

Transpose for X_C

$$\frac{230}{7.5} = X_C$$

$$\frac{230}{7.5} = 30.6\,\Omega$$

Step 2 to find C

$$X_C = \frac{10^6}{2\pi f C}$$

Transpose $C = \dfrac{10^6}{2\pi \times f \times X_C}$

$$C = \frac{10^6}{(2 \times 3.142 \times 50 \times 30.6)}$$

$\quad\quad = 104$ answer in microfarads (μF)
$\quad\quad$ (Note simply change places of X_C and C)

Enter on calculator EXP $6 \div (2\pi \times 50 \times 30.6)$ or
EXP $6 \div (2 \times 3.142 \times 50 \times 30.6)$

EXERCISE 3

1. Determine the reactance of each of the following capacitors
   to a 50 Hz supply. (Values are all in microfarads.)
   (a) 60   (d) 150   (g) 250   (j) 75
   (b) 25   (e) 8   (h) 95
   (c) 40   (f) 12   (i) 16
2. Calculate the value of capacitors which have the following
   reactances at 50 Hz. (Values are all in ohms).
   (a) 240   (d) 4.5   (g) 45   (j) 72
   (b) 75   (e) 36   (h) 400
   (c) 12   (f) 16   (i) 30
3. Calculate the value of a capacitor which will take a current
   of 25 A from a 230 V 50 Hz supply.
4. A capacitor in a certain circuit is passing a current of 0.2 A
   and the voltage drop across it is 100 V. Determine its value
   in microfarads. The frequency is 50 Hz.
5. A 20 μF capacitor is connected to an alternator whose
   output voltage remains constant at 150 V but whose
   frequency can be varied from 25 to 60 Hz. Draw graph to

show the variation in current through the capacitor as the frequency changes over this range.

6. Calculate the voltage drop across a 5 µF capacitor when a current of 0.25 A at 50 Hz flows through it.

7. In order to improve the power factor of a certain installation, a capacitor which will take 15 A from the 230 V supply is required. The frequency is 50 Hz. Calculate the value of the capacitor.

8. In one type of twin-tube fluorescent fitting, a capacitor is connected in series with one of the tubes. If the value of the capacitor is 7 µF, the current through it is 0.8 A, and the supply is at 50 Hz, determine the voltage across the capacitor.

9. A machine designed to work on a frequency of 60 Hz has a power-factor-improvement capacitor which takes 12 A from a 110 V supply. Calculate the current the capacitor will take from the 110 V 50 Hz supply.

10. A capacitor takes a current of 16 A from a 400 V supply at 50 Hz. What current will it take if the voltage falls to 380 V at the same frequency?

11. A 22 µF capacitor is connected in an a.c. circuit at 50 Hz. Its reactance is
(a) 0.000 145 Ω
(c) 6 912 000 Ω
(b) 6912 Ω
(d) 145 Ω

12. The value in microfarads of a capacitor which has a capacitive reactance of 100 Ω when connected to a circuit at 50 Hz is
(a) 31.8 µF
(c) 0.000 0318 µF
(b) 318 µF
(d) 0.0314 µF

IMPEDANCE IN SERIES CIRCUITS

When resistance (R) is in a circuit with reactance $(X_L$ or $X_C)$, the combined effect is called Impedance (Z), this is measured in ohms.

For series circuits, the calculation for impedance (Z) is

$$Z^2 = R^2 + X^2 \quad \text{or} \quad Z = \sqrt{R^2 + X^2}$$

In this calculation X is for X_C or X_L.

Where the circuit contains inductive reactance (X_C) and capacitive reactance (X_L).

$$X = X_C - X_L \quad \text{or} \quad X_L - X_C$$

X will be the largest reactance minus the smallest reactance.

An inductor coil will always possess both inductance (the magnetic part of the circuit) and resistance (the resistance of the wire), together they produce impedance. Although inductance and impedance cannot be physically separated, it is convenient for the purpose of calculation to show them separately in a circuit diagram.

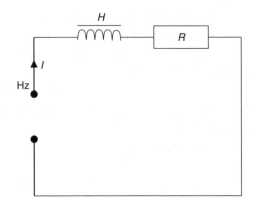

Fig. 4

EXAMPLE 1   A coil has a resistance of 6 Ω and an inductance of 0.09 H. Calculate its impedance to a 50 Hz supply.

Step 1

Inductive reactance $X_L = 2\pi f L$

$$2\pi \times f \times 0.09$$

$$2 \times 3.142 \times 50 \times 0.09 = 28.27 \, \Omega$$

(Note: a common error is to add the resistance and inductance treating it as a d.c. circuit)

Step 2

$$Z^2 = R^2 + X_L^2$$

$$\text{or } Z = \sqrt{R^2 + X^2}$$

$$Z = \sqrt{6^2 + 28.27^2}$$

$$= 29.32 \, \Omega$$

Enter into calculator $6X^2 + 28.27X^2 = \sqrt{} = $ (answer 28.9 Ω).

EXAMPLE 2   A coil passes a current of 23 A when connected to a 230 V d.c. supply, but only 8 A when connected to a 230 V supply.

When connected to a d.c. circuit the coil's resistance is only that of the wire in the coil, this can be calculated using Ohm's law. On d.c.

$$U = I \times R$$

$$\frac{U}{I} = R$$

$$\frac{230}{23} = 10 \, \Omega \,(\text{resistance})$$

On an a.c. circuit, reactance will be produced, as this is an inductive circuit it will be inductive reactance (X_L).

The combined effect of the resistance and reactance of the coil is the impedance (Z).

Step 1

$$\text{On a.c. } U = I \times Z$$

$$230 = 8 \times Z$$

Transpose

$$\frac{230}{8} = 28.75 \, \Omega \text{ impedance } (Z).$$

Step 2

To find the inductance of the coil.

$$Z^2 = R^2 + X_L^2$$
$$X_L^2 = Z^2 - R^2$$
$$X_L = \sqrt{28.7^2 - 10^2}$$
$$X_L = 26.90\,\Omega$$

Enter on calculator

$$28.7X^2 - 10^2 = \sqrt{} = (\text{answer } 26.90\,\Omega)$$

Step 3

$$X_L = 2\pi f L$$
$$26.90 = 2 \times 3.142 \times 50 \times L$$

Transpose

$$\frac{26.90}{(2 \times 3.142 \times 50)} = L = 0.085\,\text{H}$$

Enter on calculator

$$26.90 \div (2 \times 3.142 \times 50) = (\text{answer})$$

EXAMPLE 3   A 70 Ω resistor is wired in series with a capacitor of an unknown value to a 230 volt 50 Hz supply.

Calculate the value of the capacitor in microfarads if a current of 1.3 A flows.

First find impedance of circuit (Z)

Step 1

$$U = I \times Z$$
$$230 = 1.3 \times Z$$
$$Z = \frac{230}{1.3}$$
$$Z = 176.92\,\Omega$$

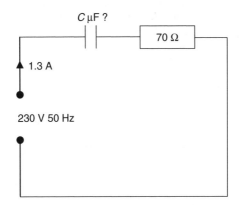

Fig. 5

Step 2
Next find capacitive reactance X_C

$$Z^2 = R^2 + X_C^2$$
$$176.92 = \sqrt{70^2 + X_C^2}$$

Transpose for X_C

$$X_C = \sqrt{176.92^2 - 70^2}$$
$$\dot{X}_C = 162.48\ \Omega$$

Now find capacitance
Step 3

$$X_C = \frac{10^6}{2\pi f C}$$

Transpose for C

$$C = \frac{10^6}{2\pi f X_L}$$
$$C = \frac{10^6}{2 \times 3.142 \times 50 \times 162.48}$$

19.59 μF is the capacitor value
On calculator enter EXP 6 ÷ (2 × 3.142 × 50 × 162.48) = (answer)

EXAMPLE 4   A coil of inductance of 0.09 H and a resistance of 18 Ω is wired in series with a 70 μF capacitor to a 230 volt 50 Hz supply.

Calculate the current which flows and the voltage drop across the capacitor.

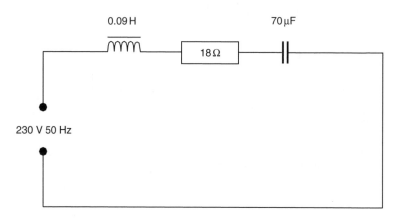

0.09 H   70 μF

18 Ω

230 V 50 Hz

Fig. 6

Step 1
Calculate inductive and capacitive reactance.

Inductive reactance

$$X_L = 2\pi f L$$
$$= 2 \times 3.142 \times 50 \times 0.09$$
$$= 28.27\ \Omega$$

Capacitive reactance

$$X_C = \frac{10^6}{2\pi f C}$$
$$= \frac{10^6}{2 \times 3.142 \times 50 \times 70}$$
$$= 45.46\ \Omega$$

Enter on calculator EXP $6 \div (2 \times 3.142 \times 50 \times 70) =$ (answer)

Step 2

Find the actual reactance for circuit which is the largest reactance minus the smallest reactance

For this circuit

$$X = X_C - X_L$$

$$= 45.46 - 28.27$$

$$= 17.19 \; \Omega \text{ (this is } X_C \text{ as the capacitive reactance is larger than the inductive reactance)}$$

Step 3

Calculate the impedance for the circuit (Z)

Impedance Z is found

$$Z^2 = R^2 + X^2$$

$$Z^2 = 18^2 + 17.19^2$$

$$Z = \sqrt{18^2 + 17.19^2}$$

Enter on calculator $18X^2 + 17.19X^2 = \sqrt{} = (\text{answer})$

$$Z = 24.88 \; \Omega$$

Step 4

Calculate current (I)

$$U = I \times Z$$

$$230 = I \times 24.88$$

Transpose for I

$$\frac{230}{24.88} = 9.24 \, \text{A}$$

As this current is common to the whole circuit, the voltage across the capacitor and the inductor can be calculated.

If a phasor is required the current is the reference conductor.

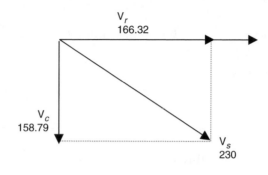

Fig. 7

Voltage across capacitor

$$U_C = I \times X_C$$
$$= 9.24 \times 45.46$$
$$= 420 \, \text{volts}$$

Voltage across inductor

$$U_I = I \times X_L$$
$$= 9.24 \times 28.27$$
$$= 261.21 \, \text{volts}$$

(Note both voltages are higher than the 230 V supply. This often happens in a.c. circuits. The voltages do not add up as in d.c. circuits.)

 EXERCISE 4

1. Complete the following table:

R	15	25	3.64				76.4	0.54
R^2				2250	18.7	40		

2. Complete the following table:

X	29.8		0.16			897		
X^2		0.46		0.9	0.16		54 637	0.036

3. A coil of wire has resistance of $8\,\Omega$ and inductance of $0.04\,\text{H}$. It is connected to supply of $100\,\text{V}$ at $50\,\text{Hz}$. Calculate the current which flows.

4. An inductor of inductance $0.075\,\text{H}$ and resistance $12\,\Omega$ is connected to a $230\,\text{V}$ supply at $50\,\text{Hz}$. Calculate the current which flows.

5. Complete the following table:

$R\,(\Omega)$	14.5		9.63	3.5	57.6	
$X\,(\Omega)$	22.8	74.6		34.7		49.6
$Z\,(\Omega)$		159	18.4		4050	107

6. A capacitor of $16\,\mu\text{F}$ and a resistor of $120\,\Omega$ are connected in series. Calculate the impedance of the circuit.

7. A resistor of $200\,\Omega$ and a capacitor of unknown value are connected to a $230\,\text{V}$ supply at $50\,\text{Hz}$ and a current of $0.85\,\text{A}$ flows. Calculate the value of the capacitor in microfarads.

8. When a certain coil is connected to a $110\,\text{V}$ d.c. supply, a current of $6.5\,\text{A}$ flows. When the coil is connected to a $110\,\text{V}$ $50\,\text{Hz}$ a.c. supply, only $1.5\,\text{A}$ flows. Calculate (a) the resistance of the coil, (b) its impedance, and (c) its reactance.

9. The inductor connected in series with a mercury vapour lamp has resistance of $2.4\,\Omega$ and impedance of $41\,\Omega$. Calculate the inductance of the inductor and the voltage drop across it when the total current in the circuit is $2.8\,\text{A}$. (Assume the supply frequency is $50\,\text{Hz}$.)

10. An inductor takes $8\,\text{A}$ when connected to a d.c. supply at $230\,\text{V}$. If the inductor is connected to an a.c. supply at $230\,\text{V}$ $50\,\text{Hz}$, the current is $4.8\,\text{A}$. Calculate (a) the resistance, (b) the inductance, and (c) the impedance of the inductor.

11. What is the function of an inductor in an alternating-current circuit?

    When a d.c. supply at $230\,\text{V}$ is applied to the ends of a certain inductor coil, the current in the coil is $20\,\text{A}$. If an a.c. supply at $230\,\text{V}$ $50\,\text{Hz}$ is applied to the coil, the current in the coil is $12.15\,\text{A}$.

Calculate the impedance, reactance, inductance, and resistance of the coil.

What would be the general effect on the current if the frequency of the a.c. supply were increased?

12. A coil having constant inductance of 0.12 H and resistance of 18 Ω is connected to an alternator which delivers 100 V a.c. at frequencies ranging from 28 to 55 Hz. Calculate the impedance of the coil when the frequency is 30, 35, 40, 45 and 50 Hz and plot a graph showing how the current through the coil varies according to the frequency.

13. The inductor in a discharge lighting circuit causes a voltage drop of 120 V when the current through it is 2.6 A.

Determine the size in microfarads of a capacitor which will produce the same voltage drop at the same current value. (Neglect the resistance of the inductor. Assume the supply frequency is 50 Hz.)

14. A circuit is made up of an inductor, a resistor and a capacitor all wired in series. When the circuit is connected to a 50 Hz a.c. supply, a current of 2.2 A flows. A voltmeter connected to each of the components in turn indicates 220 V across the inductor, 200 V across the resistor, and 180 V across the capacitor. Calculate the inductance of the inductor and the capacitance of the capacitor.

At what frequency would these two components have the same reactance value? (Neglect the resistance of the inductor.)

15. What are meant by the following terms used in connection with alternating current: resistance, impedance and reactance?

A voltage of 230 V, at a frequency of 50 Hz, is applied to the ends of a circuit containing a resistor of 5 Ω, an inductor of 0.02 H, and a capacitor of 150 μF, all in series. Calculate the current in the circuit.

16. A coil of resistance 20 Ω and inductance 0.08 H is connected to a supply at 240 V 50 Hz. Calculate (a) the current in the circuit, (b) the value of a capacitor to be put in series with the coil so that the current shall be 12 A. (CGLI)

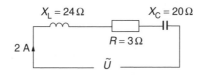

Fig. 8

17. For the circuit shown in Figure 8, the voltage V is

    (a) 94 V   (b) 14 V   (c) 10 V   (d) 0.043 V

18. An inductor has inductance 0.12 H and resistance 100 Ω. When it is connected to a 100 V supply at 150 Hz, the current through it is

    (a) 1.51 A   (b) 0.47 A   (c) 0.66 A   (d) 0.211 A

IMPEDANCE TRIANGLES AND POWER TRIANGLES

For a right-angled triangle (Figure 9), the theorem of Pythagoras states that

$$a^2 = b^2 + c^2$$

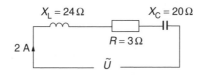

Fig. 9

As the relationship between impedance, resistance and reactance in a series circuit is given by an equation of a similar form, $Z^2 = R^2 + X^2$, conditions in such circuits can conveniently be represented by right-angled triangles. In Figure 10,

$$Z^2 = R^2 + X^2$$

where $X = X_L$ (Fig. 9(a))   or X_C (Fig. 9(b))

and ϕ = the *phase angle* of the circuit

$$\sin\phi = \frac{X}{Z} \quad \cos\phi = \frac{R}{Z} \quad \text{and} \quad \tan\phi = \frac{X}{R}$$

$\cos\phi$ is the *power factor* of the circuit.

27

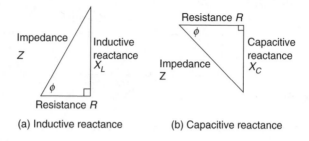

(a) Inductive reactance   (b) Capacitive reactance

Fig. 10

A right-angled triangle is also used to represent the apparent power in a circuit and its active and reactive components (Figure 11).

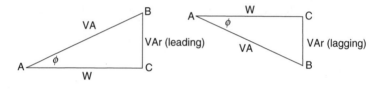

Fig. 11

AB is the product of voltage and current in the circuit (VA).
AC is the true power – the working component (W).
BC is the reactive or wattless component (VAr).

$$\frac{\text{VAr}}{\text{VA}} = \sin\phi$$

∴ $\text{VAr} = \text{VA} \times \sin\phi$

$$\frac{\text{W}}{\text{VA}} = \cos\phi$$

∴ $\text{W} = \text{VA}\cos\phi$

and $\cos\phi$ is the power factor (p.f.).

In power circuits, the following multiples of units are used:

     kVA   kW   and   kVAr

EXAMPLE I Find Z in Figure 12.

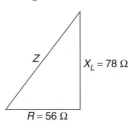

Fig. 12

$$Z^2 = R^2 + X_L^2$$
$$= 56^2 + 78^2$$
$$= 3135 + 6084$$
$$= 9219$$
$$\therefore \quad Z = \sqrt{9219}$$
$$= 96.02$$
$$= 96\ \Omega \text{ (correct to three significant figures)}$$

EXAMPLE 2 Find X_C in Figure 13.

$$Z^2 = R^2 + X_C^2$$
$$125^2 = 67.2^2 + X_C^2$$
$$\therefore \quad X_C^2 = 125^2 - 67.6^2$$
$$= 15\,620 - 4570$$
$$= 11\,050$$
$$\therefore \quad X_C = \sqrt{11\,050} = 105.1$$
$$= 105\ \Omega$$

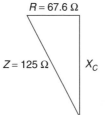

Fig. 13

Alternatively,

$$Z^2 = R^2 + X_C^2$$
$$125^2 = 67.6^2 + X_C^2$$
$$\therefore \quad X_C^2 = 125^2 - 67.6^2$$
$$= (125 - 67.6)(125 - 67.6)$$
$$= 192.6 \times 57.4$$
$$= 11\,050$$
$$\therefore \quad X_C = \sqrt{11\,050}$$
$$= 105\,\Omega$$

EXAMPLE 3   Find ϕ in Figure 14.

$$\tan\phi = \frac{X_L}{R}$$
$$= \frac{15}{20} = 0.75$$
$$\therefore \quad \phi = 36°52'$$

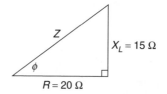

Fig. 14

EXAMPLE 4   Find X_C in Figure 15.

$$\frac{X_C}{Z} = \sin\phi$$
$$\frac{X_C}{90} = \sin 48° = 0.7431$$
$$\therefore \quad X_C = 90 \times 0.7431$$
$$= 66.9 \,(\text{to three significant figures})$$

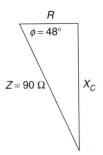

R

$\phi = 48°$

$Z = 90\ \Omega$ X_C

Fig. 15

EXAMPLE 5   Find the kVA and kVAr in Figure 16.

$$\frac{kW}{kVA} = \cos\phi$$

$$\frac{15}{kVA} = \cos 42° = 0.7431$$

$\therefore$ $$\frac{kVA}{15} = \frac{1}{0.7431}$$

$\therefore$ $$kVA = \frac{15}{0.7431}$$

$$= 20.2$$

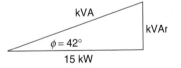

kVA

kVAr

$\phi = 42°$

15 kW

Fig. 16

$$\frac{kVAr}{kW} = \tan\phi$$

$\therefore$ $$\frac{kVAr}{15} = \tan 42° = 0.9004$$

$\therefore$ $$kVAr = 15 \times 0.9004$$

$$= 13.5$$

EXAMPLE 6 A coil of 0.2 H inductance and negligible resistance is connected in series with a 50 Ω resistor to the 230 V 50 Hz mains (Figure 17). Calculate (a) the current which flows, (b) the power factor, (c) the phase angle between the current and the applied voltage.

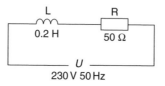

Fig. 17

Coil reactance $X_L = 2\pi f L$

$$= 2\pi \times 50 \times 0.2$$
$$= 314 \times 0.2$$
$$= 62.8\,\Omega$$

To find the impedance (Figure 18),

$$Z^2 = R^2 + X_L^2$$
$$= 50^2 + 62.8^2$$
$$= 2500 + 3944$$
$$= 6444$$

∴ $$Z = \sqrt{6444}$$
$$= 80.27\,\Omega$$

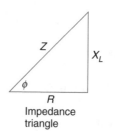

Impedance triangle

Fig. 18

(a) To find the current,

$$U = I \times Z$$

$$\therefore \qquad 230 = I \times 80.27$$

$$\therefore \qquad I = \frac{230}{80.27}$$

$$= 2.86\,\text{A}$$

(b) Power factor $= \cos\phi = \dfrac{R}{Z}$

$$= \frac{50}{80.27}$$

$$= 0.623\,\text{lag}$$

(c) The phase angle is the angle whose cosine is 0.623,

$$\therefore \qquad \phi = 51°\ 28'$$

EXERCISE 5

1. Find Z in Figure 19.

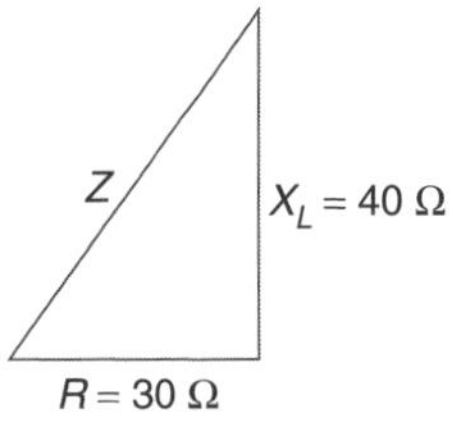

2. Find Z in Figure 20.

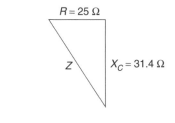

Fig. 19

Fig. 20

3. Find R in Figure 21.

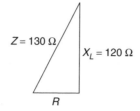

Fig. 21

4. Find X_C in Figure 22.

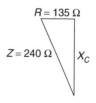

Fig. 22

5. Find R in Figure 23.

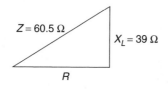

Fig. 23

6. Find Z in Figure 24.

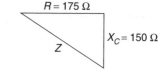

Fig. 24

7. Find R in Figure 25.

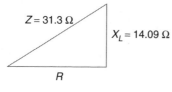

Fig. 25

8. Find X_L in Figure 26.

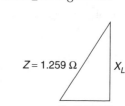

Fig. 26

9. Find Z in Figure 27.

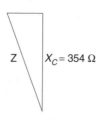

Fig. 27

10. Find X_L in Figure 28.

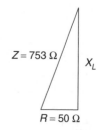

Fig. 28

11. Find R in Figure 29.

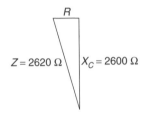

Fig. 29

12. Consider the answers to questions 9 to 11 and then write down the approximate impedance of a coil which has resistance 32 Ω and reactance 500 Ω.

13. Complete the following table:

Angle ϕ 30° 45° 60° 90° 52° 24′ 26° 42′ 83° 12′ 5° 36′
sin ϕ
cos ϕ
tan ϕ

14. Complete the following table:

Angle ϕ 33° 3′ 75° 21′ 17° 15′ 64° 29′ 27° 56′ 41° 53′
sin ϕ
cos ϕ
tan ϕ

15. Complete the following table:

Angle ϕ						
sin ϕ			0.91	0.6		0.9088	
cos ϕ		0.9003		0.8			0.4754
tan ϕ	0.4000				1.2088		

16. Complete the following table:

Angle ϕ			38° 34′			
sin ϕ	0.9661					
cos ϕ		0.4341		0.8692	0.3020	0.318
tan ϕ			0.0950		3.15	

17. Find R and X_L in Figure 30.

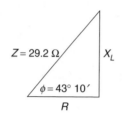

Fig. 30

18. Find R and X_C in Figure 31.

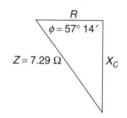

Fig. 31

19. Find ϕ in Figure 32.

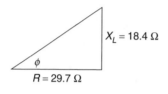

Fig. 32

20. Calculate Z and X_L in Figure 33.

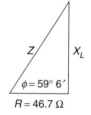

Fig. 33

21. Find W and VAr in Figure 34.

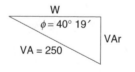

Fig. 34

22. Find ϕ and X_L in Figure 35.

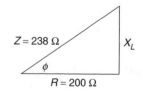

Fig. 35

21. Find ϕ in Figure 36.   **22.** Calculate R in Figure 37.

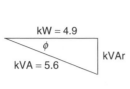

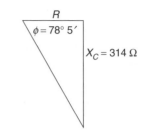

Fig. 36   **Fig. 37**

23. Find OX in Figure 38.   **24.** Find OX in Figure 39.

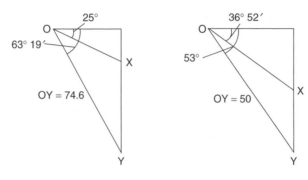

Fig. 38   **Fig. 39**

25. Complete the following table then plot a graph of power factor ($\cos \phi$) to a base of phase angle (ϕ):

Phase angle ϕ		65° 6′ 60°		45° 40′
Power factor $\cos \phi$	0.25 0.3		0.55 0.6	0.82

26. A coil has inductance 0.18 H and resistance 35 Ω. It is connected to a 100 V 50 Hz supply. Calculate (a) the impedance of the coil, (b) the current which flows, (c) the power factor, (d) the power absorbed by the coil.

27. Define the term 'power factor' and state how it affects cable size.

An inductor of resistance 8 Ω and of inductance 0.015 H is connected to an alternating-current supply at 230 V, single-phase, 50 Hz. Calculate (a) the current from the supply, (b) the power in the circuit, (c) the power factor.

28. A single-phase alternating-current supply at 230 V 50 Hz is applied to a series circuit consisting of an inductive coil of negligible resistance and a non-inductive resistance coil of 15 Ω. When a voltmeter is applied to the ends of each coil in turn, the potential differences are found to be 127.5 V across the inductive coil, 203 V across the resistance.

   Calculate (a) the impedance of the circuit, (b) the inductance of the coil, (c) the current in the circuit, and (d) the power factor. (CGLI)

29. On what factors do the resistance, reactance and impedance of an alternating-current circuit depend, and how are these quantities related?

   The current in a single-phase circuit lags behind the voltage by 60°. The power in the circuit is 3600 W and the voltage is 240 V. Calculate the value in ohms of the resistance, the reactance and the impedance. (CGLI)

a.c. waveform and phasor representation

ALTERNATING E.M.F. AND CURRENT

The value and direction of the e.m.f. induced in a conductor
rotating at constant speed in a uniform magnetic field,
Figure 40(a) vary according to the position of the conductor.

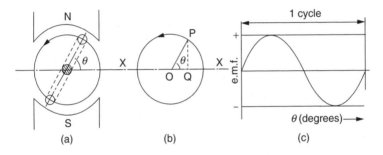

Fig. 40

The e.m.f. can be represented by the displacement QP of the
point P above the axis XOX, Figure 40(b). OP is a line which is
rotating about the point O at the same speed as the conductor is
rotating in the magnetic field. The length of OP represents the
maximum value of the induced voltage. OP is called a *phasor*.

A graph, Figure 40(c), of the displacement of the point P
plotted against the angle θ (the angle through which the
conductor has moved from the position of zero induced e.m.f.) is
called a *sine* wave, since the PQ is proportional to the sine angle θ.
One complete revolution of OP is called a *cycle*.

EXAMPLE I   An alternating voltage has a maximum value of
200 V. Assuming that it is sinusoidal in nature (i.e. it varies

according to a sine wave), plot a graph to show the variations in this voltage over a complete cycle.

Method (Figure 41) Choose a reasonable scale for OP; for instance, 10 mm ≡ 100 V.

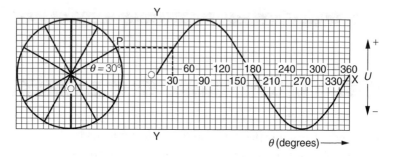

Fig. 41

Draw a circle of radius 20 mm at the left-hand side of a piece of graph paper to represent the rotation of OP.

One complete revolution of OP sweeps out 360°. Divide the circle into any number of equal portions, say 12. Each portion will then cover 30°.

Construct the axes of the graph, drawing the horizontal axis OX (the *x*-axis) on a line through the centre of the circle. This *x*-axis should now be marked off in steps of 30° up to 360°. If desired, perpendicular lines can be drawn through these points. Such lines are called *ordinates*.

The points on the graph are obtained by projecting from the various positions of P to the coordinate corresponding to the angle θ at that position.

Remember that when $\theta = 0°$ and $180°$ the generated e.m.f. is zero, and when $\theta = 90°$ and $270°$ the generated e.m.f. has its maximum value.

EXAMPLE 2   Two alternating voltages act in a circuit. One (A) has an r.m.s. value of 90 V and the other (B) has an r.m.s. value of 40 V, and A leads B by 80°. Assuming that both voltages are sinusoidal, plot graphs to show their variations over a complete cycle. By adding their instantaneous values together, derive a graph of the resultant voltage. Give the r.m.s. value of this resultant.

First find the maximum values of the voltages given:

$$U_{\text{r.m.s.}} = 0.707 \times U_{\text{max}}$$

$$\therefore \qquad 90 = 0.707 \times U_{\text{max}}$$

$$\therefore \qquad U_{\text{max}} = \frac{90}{0.707}$$

$$= 127\,\text{V}$$

Similarly, if

$$U_{\text{r.m.s.}} = 40$$

$$U_{\text{max}} = \frac{40}{0.707}$$

$$= 56.6\,\text{V}$$

Choose a suitable scale, say 20 mm ≡ 100 V. Draw two circles with the same centre, one having a radius of 25.4 mm (127 V), the other a radius of 11.32 mm (56.6 V).

Draw phasors to represent the voltages: OA horizontal and OB, which represents the lower voltage, lagging 80° behind OA (anticlockwise rotation is always used) – see Figure 42.

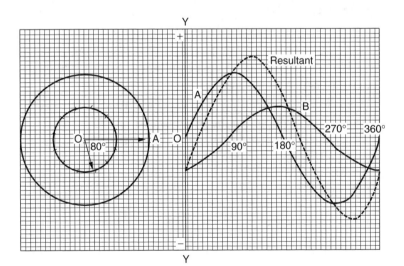

Fig. 42

Mark off the circumference of the larger circle in steps of 30°, using OA as the reference line.

Mark off the smaller circle in steps of 30°, using OB as the reference line.

Set off the axes of the graph alongside as in the previous example.

Plot the sine wave of voltage A as before.

Plot the sine wave of voltage B in exactly the same way, projecting the first point from B to the y-axis YOY and from each succeeding 30° point to the appropriate 30° point on the horizontal axis of the graph.

Points on the resultant graph are plotted by combining the ordinates of A and B at each 30° point. If the graphs lie on the same side of the x-axis, the ordinates are added. If the graphs lie on opposite sides of the axis, the smaller is subtracted from the larger (measurements upwards from the x-axis are positive, measurements downwards are negative).

The resultant curve is shown by the dotted line in Figure 42 and its maximum value is approximately 150 V.

Its r.m.s. value is

$$0.707 \times 150 = 106\,\text{V}$$

EXAMPLE 3 A current of 15 A flows from the 230 V mains at a power factor of 0.76 lagging. Assuming that both current and voltage are sinusoidal, plot graphs to represent them over one cycle. Plot also on the same axes a graph showing the variation in power supplied over one cycle.

The procedure for plotting the current and voltage sine waves is the same as that adopted in the previous example.

The phase angle between current and voltage is found from the power factor as follows:

$$\text{power factor} = \cos\phi$$

where ϕ is the angle of phase difference

$$\cos\phi = 0.76$$

∴ $$\phi = 40°\,32'$$

$$U_{max} = \frac{230}{0.707}$$

$$= 325.3\,\text{V}$$

$$I_{max} = \frac{15}{0.707}$$

$$= 21.21\,\text{A}$$

Scales of $20\,\text{mm} \equiv 200\,\text{V}$ and $20\,\text{mm} \equiv 20\,\text{A}$ will be suitable.

To obtain the graph of the power supplied, the ordinates of current and voltage are multiplied together (Figure 43). It is convenient to do this every $30°$ as before.

Remember the rules for multiplying positive and negative numbers.

Where the resulting graph is negative, additional points are helpful in obtaining a smooth curve.

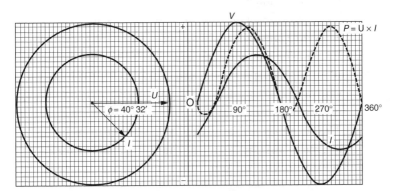

Fig. 43

That portion of the power curve lying above the x-axis represents the power supplied to the circuit. That portion lying below the x-axis represents the power returned to the mains from the circuit.

1. Plot a sine wave, over one complete cycle, of an alternating voltage having a maximum value of 325 V. Determine the r.m.s. value of this voltage.

2. An alternating current has the following value taken at intervals of 30° over one half cycle:

Angle ϕ	0	30°	60°	90°	120°	150°	180°
Current (A)	0	10.5	17.5	19.7	15.0	11.5	0

   Determine the average and r.m.s. values of this current.

3. Plot a graph over one complete cycle of a sinusoidal alternating voltage having an r.m.s. value of 200 V.

4. Two sinusoidal voltages act in a circuit. Their r.m.s. values are 110 V and 80 V and they are out of phase by 75°, the lower voltage lagging. Plot sine waves on the same axes to represent these voltages. Plot a graph of the resultant voltage by adding together the ordinates of the two waves. Give the r.m.s. value of the resultant voltage and state approximately the phase angle between this resultant and the lower voltage.

5. Two alternating currents are led into the same conductor. They are sinusoidal and have r.m.s. values of 4 A and 1 A. The smaller current leads by 120°. Plot out the sine waves of these two currents and add the ordinates to obtain the sine wave of the resultant current. Calculate the r.m.s. value of the resultant.

6. The current taken by an immersion heater from the 250 V a.c. mains is 12.5 A. Current and voltage are in phase and are sinusoidal. Plot graphs on the same axes to show the variations in current and voltage over one complete cycle.

7. A 10 µF capacitor is connected to a 240 V supply at 50 Hz. The current leads the voltage by 90°, and both may be assumed to be sinusoidal. Plot the sine waves of the current and voltage over one complete cycle.

8. A fluorescent lamp takes a current of 1.2 A from a 230 V supply at a power factor of 0.47. Assuming that both

current and voltage are sinusoidal, plot graphs to show how they vary over a complete cycle.

9. The current in a circuit is 25 A and the supply voltage is 220 V. The power factor is 0.6 lagging. Plot sine waves to represent current and voltage over one cycle. Multiply the instantaneous values of current and voltage together to obtain a graph representing the power in the circuit.

10. An inductor of 0.1 H is connected to a 100 V supply at 50 Hz. Neglecting the resistance of the winding, calculate the current which flows. Plot sine waves to represent the current and voltage, assuming that the voltage leads the current by 90°. Multiply the ordinates of the waves together to obtain a graph representing the power supplied to the circuit.

PHASORS

Conditions is alternating-current circuits can be represented by means of phasor diagrams.

In Figure 44, U is a voltage and I is a current, ϕ is the angle of phase difference, and $\cos \phi$ is the power factor.

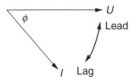
(a) Lagging power factor

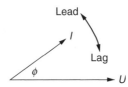
(b) Leading power factor

Fig. 44

EXAMPLE I The current in a circuit is 5 A, the supply voltage is 230 V, and the power factor is 0.8 lagging. Represent these conditions by means of a phasor diagram drawn to scale. Choose a suitable scale.

Power factor $= 0.8$

$$= \cos \phi$$

$$\cos \phi = 0.8$$

$$\phi = 36° \, 52' \quad \text{(see Figure 45)}$$

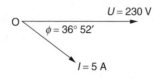

Fig. 45

Normally the r.m.s. values are used when drawing phasor diagrams.

Note that the most accurate construction is obtained by setting off two lines at the required angle and then marking the lines to the appropriate lengths from the point of intersection with compasses which have been set to the respective measurement.

EXAMPLE 2 A resistor and a capacitor are wired in series to an a.c. supply (Figure 46). When a voltmeter is connected across the resistor it reads 150 V. When it is connected to the capacitor terminals it indicates 200 V. Draw the phasor diagram for this circuit to scale and thus determine the supply voltage.

As the value of current is not given, it will not be possible to draw its phasor to scale.

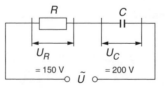

Fig. 46

The current is the same throughout a series circuit and so the current phasor is used as a reference.

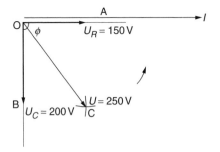

Fig. 47

Draw OI any length to represent the current (Figure 47).

From point O, draw thin lines parallel to and at right angles to OI (capacitor voltage *lags* behind the current).

Choose a suitable scale and use compasses set to the required measurement to mark off OA = U_R, the resistor voltage drop – in phase with the current – and OB = U_C, the capacitor voltage drop.

With centre A and compasses set to length OB, strike an arc. With centre B and compasses set to OA, strike another arc. These arcs intersect at point C.

OC is the resultant voltage, which is equal to the supply voltage.

By measurement of OC, the supply voltage is found to be 250 V.

EXAMPLE 3 An inductor takes a current of 5 A from a 230 V supply at a power factor of 0.4 lagging. Construct the phasor diagram accurately to scale and estimate from the diagram the resistance and reactance of the coil.

As already explained, although resistance and reactance cannot be separated, it is convenient to draw them apart in an equivalent circuit diagram (Figure 48). The total voltage drop – in this case the supply voltage – will then be seen to be made up of a resistance voltage drop and a reactance voltage drop.

Since, again, we are considering a series circuit in which the current is the same throughout, it is not necessary to draw the current phasor to scale.

Power factor $= \cos\phi$

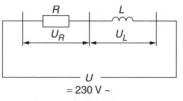

$$U$$
$$= 230 \text{ V} \sim$$

Equivalent circuit diagram

Fig. 48

where ϕ is the angle of phase difference between current and supply voltage

and $\cos \phi = 0.4$

$\therefore \qquad \phi = 66° \, 25'$

Draw OI any length to represent the current (Figure 49).

Choose a suitable scale and set off OC at 66° 25′ from OI and of length to represent the supply voltage.

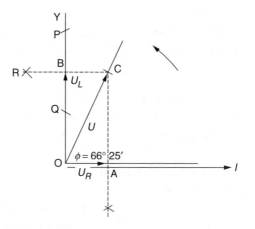

Fig. 49

Draw OY at right angles to the current phasor and from C draw perpendiculars to cut the current phasor at A and OY at B. The perpendiculars are constructed as follows:

(i) Set the compasses to any radius and with centre C draw arcs which cut OY at P and Q.

(ii) With the compasses again set to any radius and with centres P and Q strike two more arcs to cut in R. CR is then perpendicular to OY.

A similar method is employed in drawing CA.

By measurement,

$$U_R = 93\,\text{V}$$

$$U_L = 209\,\text{V}$$

Now $\quad U_R = I \times R$

$\therefore \quad\quad 93 = 5 \times R$

$\therefore \quad\quad R = \dfrac{93}{5}$

$$= 18.5\,\Omega$$

and $\quad U_L = I \times X_L$ (X_L is the inductive reactance)

$\therefore \quad\quad 209 = 5 \times X_L$

$\therefore \quad\quad X_L = \dfrac{209}{5}$

$$= 41.8\,\Omega$$

EXAMPLE 4 An appliance takes a single-phase current of 32 A at 0.6 p.f. lagging from a 250 V a.c. supply. A capacitor which takes 8.9 A is wired in parallel with this appliance (Figure 50). Determine graphically the total supply current.

As this is a parallel circuit, the voltage is common to both branches and is thus used as the reference phasor. It need not be drawn to scale.

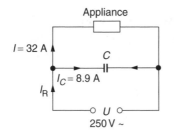

Fig. 50

Choose a suitable scale.

$$\text{p.f.} = \cos\phi = 0.6$$

$$\therefore \qquad \phi = 53°\,8'$$

Draw the voltage phasor (Figure 51) and set off the appliance-current phasor at 53° 8′ lagging (OA).

The capacitor current, 8.9 A, leads on the voltage by 90° and is drawn next (OB).

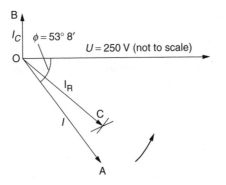

Fig. 51

The resultant of these two phasors is found as follows:
(i) With compasses set to OA and centre B, strike an arc.
(ii) With centre A and compasses set to OB, strike another arc cutting the first in C.

OC is the resultant current. By measurement of OC, the resultant current is 25.5 A.

EXAMPLE 5   A consumer's load is 15 kVA single-phase a.c. at 0.8 power factor lagging. By graphical construction, estimate the active and reactive components of this load.

$$\text{p.f.} = \cos\phi = 0.8$$

$$\therefore \qquad \phi = 36°\,52'$$

Choose a suitable scale.

Draw a thin horizontal line OX (Figure 52). Set off OA to represent 15 kVA at an angle of 36° 52′ from OX.

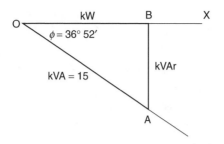

Fig. 52

From A, draw a perpendicular to cut line OX at B. OB is then the working or active component and AB is the reactive or wattless component.

By measurement of OB the true power is 12 kW, and by measurement of AB the wattless component is 9 kVAr.

EXERCISE 7

1. An a.c. circuit takes a current of 15 A at a power factor of 0.75 lagging from the 230 V mains. Construct, to scale, the phasor diagram for this circuit.
2. A power-factor-improvement capacitor takes a current of 1.6 A from a 230 V supply. Draw the phasor diagram to scale.
3. A single-phase a.c. motor takes a current of 2.75 A at a power factor of 0.18 lagging when it is running on no load. On full load it takes 4.3 A at a power factor of 0.48 lagging. The supply voltage is in each case 230 V. Draw a phasor diagram to represent the no-load and full-load circuit conditions.
4. A mercury-vapour-lamp circuit takes a current of 2.8 A at a power factor of 0.45 lagging if it is used without its p.f. improvement capacitor. When the p.f. improvement

capacitor is connected, the current falls to 1.8 A at 0.7 p.f. lagging. Construct the phasor diagram to scale.

5. A capacitor is wired in series with a resistor to an a.c. supply. When a voltmeter is connected to the capacitor terminals it indicates 180 V. When it is connected across the resistor it reads 170 V. Construct the phasor diagram for this circuit accurately to scale and from it determine the supply voltage.

6. An inductor has resistance 10 Ω and when it is connected to a 240 V a.c. supply a current of 12 A flows. Draw the phasor diagram to scale.

7. A contactor coil takes a current of 0.085 A from a 250 V supply at a power factor of 0.35 lagging. Draw the phasor diagram accurately to scale and use it to determine the resistance and reactance of the coil.

8. A single-phase transformer supplies 10 kVA at 0.7 p.f. lagging. Determine by graphical construction the active and reactive components of this load.

9. The true power input to a single-phase motor is 1150 W and the power factor is 0.54 lagging. Determine graphically the apparent power input to the machine.

10. A fluorescent-lamp circuit takes a current of 1.2 A at 0.65 p.f. lagging from the 230 V a.c. mains. Determine graphically the true power input to the circuit.

11. A single-phase motor takes 8.5 A from a 230 V supply at 0.4 p.f. lagging. A capacitor which takes 4 A is connected in parallel with the motor. From a phasor diagram drawn accurately to scale, determine the resultant supply current.

12. A discharge lighting fitting takes a current of 5.2 A at 0.46 p.f. lagging when it is used without its power-factor-improvement capacitor. When this capacitor is connected the current falls to 3.2 A, the supply voltage remaining constant at 240 V. Draw the phasor diagram to represent the conditions with and without the capacitor and from it determine the current taken by the capacitor.

(Remember that the working component of the supply current is constant.)

13. A series circuit is made up of a resistor, an inductor of negligible resistance, and a capacitor. The circuit is connected to a source of alternating current, and a voltmeter connected to the terminals of each component in turn indicates 180 V, 225 V and 146 V, respectively. Construct the phasor diagram for this circuit accurately to scale and hence determine the supply voltage.

COLEG MORGANNWG
LRC
NANGARW

Parallel circuits involving resistance, inductance and capacitance

Consider a circuit with inductance and capacitance in parallel (Figure 53).

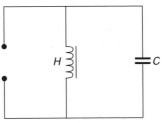

Fig. 53

where L is pure inductance (henry) and C is pure capacitance (microfarad).

In a parallel circuit the voltage is common to each branch of the circuit.

The current through the inductive branch is

$$I_L = \frac{U}{X_L}$$

where $X_L = 2\pi f L$.

This current lags the voltage by $90°$.

The current through the capacitive branch is

$$I_C = \frac{U}{X_C}$$

where $X_C = \dfrac{10^6}{2\pi f C}$ the current leads the voltage by $90°$.

Voltage is the reference and a current phasor is needed.

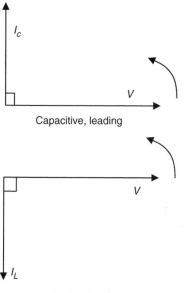

Capacitive, leading

Inductive, lagging

Fig. 54

EXAMPLE 1 Calculate the current drawn from the supply when an inductor with a reactance of 83 Ω and a capacitor of 125 Ω are connected in parallel to a 110 V supply.

Capacitor current

$$I_C = \frac{U}{X_C}$$

$$= \frac{100}{125}$$

$$= 0.8\,\text{A}$$

Inductor current

$$I_L = \frac{U}{X_L}$$

$$= \frac{100}{83}$$

$$= 1.2\,\text{A}$$

Because inductor current is larger overall, the circuit is a lagging one.

The supply current is calculated

$$I_L - I_C$$

$$= 1.2 - 0.8$$

$$= 0.4\,\text{A}$$

Lagging the voltage by $90°$

EXAMPLE 2   Calculate the current drawn from the supply when a capacitor of 75 µF is connected in parallel with a resistor of $70\,\Omega$ to a 110 volt 50 Hz supply.

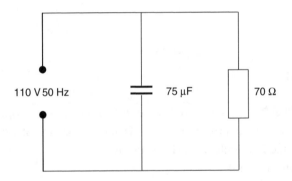

110 V 50 Hz   75 µF   70 Ω

Fig. 55

Draw a phasor diagram and determine the phase relationship between the supply voltage and the current drawn from the supply.

$$X_C = \frac{10^6}{2\pi f C}$$

$$= \frac{10^6}{2\pi \times 50 \times 75}$$

$$= 42.44\,\Omega$$

Enter into calculator EXP $6 \div (2 \text{ shift } \pi \times 50 \times 75) = $ (answer)

$$I_C = \frac{110}{42.44}$$

$$= 2.59 \, \text{A}$$

$$I_R = \frac{110}{70}$$

$$= 1.57 \, \text{A}$$

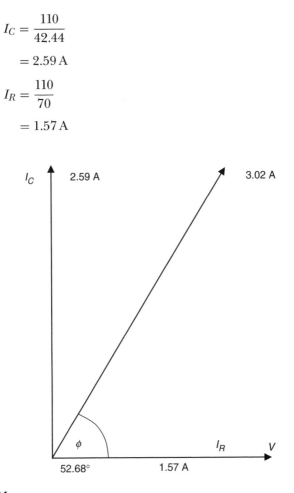

Fig. 56

Find supply current by calculation

$$I_S{}^2 = I_C^2 + I_R^2$$

$$= I_S = \sqrt{I_C + I_R}$$

$$= \sqrt{2.59 + 1.57}$$

$$= 3.02 \, \text{A}$$

Enter into calculator $2.59X^2 + 1.57X^2 = \sqrt{} = $ (answer)

To find phase angle by calculation

$$\phi = \frac{I_R}{I_S}$$

$$= \frac{1.57}{3.02}$$

$$= 0.52$$

$$\phi = 58.7°$$

Enter into calculator $1.57 \div 3.02 = $ shift $\cos^- = $ (answer)

The current is leading the supply voltage by 58.7°.

EXAMPLE 3 A coil has a resistance of 25 Ω and an inductive reactance of 20 Ω. It is connected in parallel with a capacitor of 40 Ω reactance to a 230 volt supply. Calculate the supply current and the overall power factor.

The coil impedance Z_L is

$$Z_L = \sqrt{R^2 = X_L^2}$$

$$= 25^2 + 20^2$$

$$= 32.02 \ \Omega$$

Coil current

$$I_L = \frac{U}{Z_L}$$

$$= \frac{230}{32.02}$$

$$= 7.183 \ A$$

Capacitor current

$$I_C = \frac{U}{X_C}$$

$$= \frac{230}{40}$$

$$= 5.75\,\text{A}$$

Phase angle may be calculated

$$\cos\theta = \frac{R}{Z_L}$$

$$= \frac{25}{32.02}$$

$$= 0.78$$

$$\cos\theta = 0.78$$

This is the power factor of the coil alone and is lagging.

To find phase angle, enter into calculator: shift $\cos^- 0.78 =$ (answer 38.7°)

Horizontal component of the coil current is

$$I_L \times \cos\phi = 7.2 \times 0.78 = 5.61$$

$$\text{Vertical component of coil} = \sqrt{7.2^2 - 5.61^2}$$

$$= 4.51\,\text{A}$$

Enter on calculator $7.2^2 - 5.6^2 = \sqrt{} = $ (answer)

Vertical component of capacitor current $= 5.75\,\text{A}$

$$\text{Total vertical current} = \text{capacitor current} - \text{coil current}$$

$$= 5.75 - 4.51$$

$$= 1.24\,\text{A}$$

$$I_S = \sqrt{5.62^2 + 1.24^2}$$

$$= 5.75\,\text{A}$$

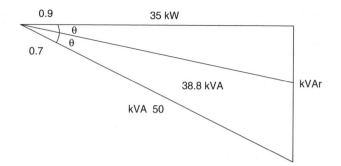

Fig. 57

---

1. Determine the current *I* in Figure 58 and state whether it leads or lags the voltage *U*.

2. Determine the resultant current *I* and its phase relationship with the supply voltage *U* in Figure 59. What is the power factor of the circuit?

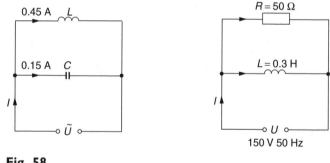

Fig. 58

Fig. 59

3. A capacitor of 15 μF is connected in parallel with a coil of inductance 0.3 H and negligible resistance to a sinusoidal supply of 240 V 50 Hz. Calculate the resultant current and state whether the phase angle is a leading or lagging one.

4. Calculate the resulting supply current and the overall power factor when a resistor of 100 Ω is connected in parallel with the circuit of question 3.

5. A coil of reactance 30 Ω and resistance 40 Ω is connected in parallel with a capacitor of reactance 200 Ω, and the circuit is supplied at 200 V. Calculate the resultant current and power factor. Check the results by constructing the phasor diagram accurately to scale.

6. A coil has resistance 150 Ω and inductance 0.478 H. Calculate the value of a capacitor which when connected in parallel with this coil to a 50 Hz supply will cause the resultant supply current to be in phase with the voltage.

7. An inductor coil of resistance 50 Ω takes a current of 1 A when connected in series with a capacitor of 31.8 μF to a 240 V 50 Hz supply. Calculate the resultant supply current when the capacitor is connected in parallel with the coil to the same supply.

8. The resultant current *I* in Figure 60 is
   (a) 0.585 A   (b) 0.085 A   (c) 11.2 A   (d) 171 A

9. The resultant current *I* in Figure 61 is
   (a) 4 A   (b) 8.5 A   (c) 2.92 A   (d) 9.22 A

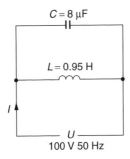

Fig. 60

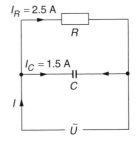

Fig. 61

Power factor improvement

EXAMPLE I   A consumer takes a load of 50 kVA at 0.7 power factor lagging. Calculate (a) the active and reactive components of the load, (b) the leading kVAr taken from a capacitor to improve the power factor to 0.9 lagging.

(a) Active component (true power) $\dfrac{kW}{kVA} = pf$

Transposed to find kW

$$kW \times 0.7 = 35\,kW$$

Reactive component $kW^2 = kVA^2 = kVAr^2$

$$\text{or} \quad kVAr = \sqrt{kW^2 - kVA^2}$$
$$= 50^2 - 35^2$$
$$= 35.7\,kVAr$$

(b) Leading kVAr required

$$\frac{kW}{kVA} = pf$$

Transposed for kVA

$$\frac{kW}{0.9} = 38.8$$
$$= \frac{35}{0.9} = 38.8$$
$$kVAr = \sqrt{38.88^2 - 35^2}$$
$$= 16.93\,kVAr$$

Lagging kVAr − leading kVAr = kVAr taken by capacitor = 35.7 − 16.93 = 18.77 kVAr

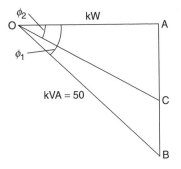

Fig. 62

EXAMPLE 2   A test on an 80 W fluorescent lamp circuit gave the following results when connected to a 50 Hz mains supply.

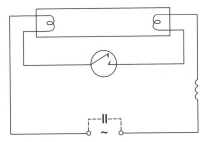

Fig. 63

Without power factor improvement capacitor

Volts 232

Amperes 1.13

Watts 122

With power factor correction capacitor

Volts 232

Amperes 0.68

Watts 122

Calculate the value of the power factor correction capacitor in microfarads (µF).

The in phase current of the circuit is calculated

$$I = \frac{P}{U}$$

$$= \frac{122}{232}$$

$$= 0.525 \, \text{A}$$

This current is common to both cases since watts are the same.

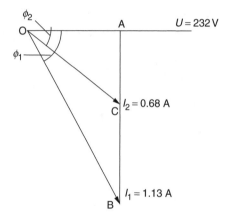

Fig. 64

CALCULATION WITHOUT P.F. CORRECTION

Current drawn from supply $= 1.13 \, \text{A}$

Wattless current in the reactive component

$$= \sqrt{1.13^2 - 0.525^2}$$

$$= 1 \, \text{A}$$

Enter into calculator $1.132 \, X^2 - 0.525 \, X^2 = \sqrt{} = $ (answer)

Current drawn from supply $= 0.68\,\text{A}$.

Wattless current in reactive component

$$= \sqrt{0.68 - 0.525}$$

$$= 0.432\,\text{A}$$

Difference between wattless current in circuit without capacitor and circuit with power factor correction capacitor:

$$= 1\,\text{A} - 0.432\,\text{A}$$

$$= 0.568\,\text{A}$$

Calculate reactance of capacitor

$$U = I \times X_C$$

$$232 = 0.568 \times X_C$$

Transpose for X_C

$$\frac{232}{0.568} = 408\,\Omega$$

$$X_C = 408\,\Omega$$

For capacitance in microfarads

$$X_C = \frac{10^6}{2\pi f C}$$

Transpose for C

$$C = \frac{10^6}{2\pi f X_C}$$

$$= \frac{10^6}{2 \times 3.142 \times 50 \times 408}$$

Enter into calculator EXP 6 $\div$ (2 shift π × 50 × 408) = (answer)

$$= 7.8\,\mu\text{F}$$

1. The nameplate of a single-phase transformer gives its rating as 5 kVA at 230 V. What is the full-load current that this transformer can supply and what is its power output when the load power factor is (a) 0.8, (b) 0.6?

2. (a) What is meant by power factor?
   (b) The installation in a factory carries the following loads: lighting 50 kW, heating 30 kW, and power 44 760 W. Assuming that the lighting and heating loads are non-inductive, and the power has an overall efficiency of 87% at a power factor of 0.7 lagging, calculate (i) the total loading in kW, (ii) the kVA demand at full load. (CGLI)

3. The current taken by a 230 V 50 Hz, single-phase induction motor running at full load is 39 A at 0.75 power factor lagging. Calculate the intake taken from the supply (a) in kW, (b) in kVA.
   Find what size capacitor connected across the motor terminals would cause the intake in kVA to be equal to the power in kW. (CGLI)

4. A group of single-phase motors takes 50 A at 0.4 power factor lagging from a 230 V supply. Calculate the apparent power and the true power input to the motors. Determine also the leading kVAr to be taken by a capacitor in order to improve the power factor to 0.8 lagging.

5. A welding set takes 60 A from a 230 V a.c. supply at 0.5 p.f. lagging. Calculate its input in (a) kVA, (b) kW.
   Determine the kVAr rating of a capacitor which will improve the power factor to 0.9 lagging. What total current will now flow?

6. Explain with the aid of a phasor diagram the meaning of power factor in the alternating-current circuit. Why is a low power factor undesirable?
   A single-phase load of 20 kW at a power factor of 0.72 is supplied at 240 V a.c. Calculate the decrease in current if the power factor is changed to 0.95 with the same kW loading. (CGLI)

7. An induction motor takes 13 A from the 240 V single-phase 50 Hz a.c. mains at 0.35 p.f. lagging. Determine the value of the capacitor in microfarads which, when connected in parallel with the motor, will improve the power factor to 0.85 lagging. Find also the supply current at the new power factor.

8. A consumer's load is 100 kVA at 0.6 p.f. lagging from a 240 V 50 Hz supply. Calculate the value of capacitance required to improve the power factor as shown in the table below:

   Power factor   0.7 0.75 0.8 0.85 0.9 0.95 1.0
   Capacitance required (µF)

9. An appliance takes a current of 45 A at 0.2 power factor lagging. Determine the current to be taken by a bank of capacitors in order to improve the power factor to 0.6 lagging. Calculate the value of the capacitors in microfarads if they are supplied at (a) 240 V, (b) 415 V, and the supply frequency is 50 Hz.

10. A test on a mercury vapour lamp gave the following results:
    Without power-factor-improvement capacitor:
    volts 230   amperes 2.22   watts 260
    With power-factor-improvement capacitor:
    volts 230   amperes 1.4   watts 260
    The supply frequency was 50 Hz. Calculate the value of the capacitor in microfarads.

11. A transformer is rated at 10 kVA 230 V. The greatest current it can supply at 0.8 p.f. is
    (a) 43.3 A   (b) 34.8 A   (c) 23 A   (d) 230 A

12. The power output of the transformer of question 11 at 0.8 p.f. is
    (a) 8 kW   (b) 12.5 kW   (c) 19.2 kW   (d) 3 kW

13. A single-phase circuit supplies a load of 20 kVA at 0.8 p.f. lagging. The kVAr rating of a capacitor to improve the power factor to unity is
    (a) 16   (b) 12   (c) 25   (d) 33.3

14. In order to improve the power factor, a circuit requires a capacitor to provide 6 kVAr at 230 V 50 Hz. Its value in microfarads is

(a) 1430 μF

(b) 143 μF

(c) 346 μF

(d) 3460 μF

Three-phase circuit calculations

Three-phase supplies to an installation are normally in-star formation with an earthed star point. The earthed star point provides a zero potential within the system to give a single phase facility, as shown in Figure 65.

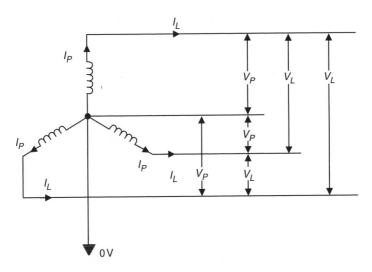

Fig. 65

The colour code and sequence for phases is L1 brown, L2 black, L3 grey.

On a standard installation, the voltage between any two phases is 400 volts, this is called the line voltage U_L, and between any phase and neutral the voltage will be 231 volts.

Calculation is

$$U_P = \frac{U_L}{\sqrt{3}}$$

$$= \frac{400}{\sqrt{3}}$$

$$= 231 \text{ volts}$$

The phasor for a balanced three-phase system is as Figure 66(b). The current in a three-phase-star connected system is $I_L = I_P$. As shown in Figure 65. I_L is the current in any line and I_P is the current phase or load.

If the currents on a star-connected supply are the same on each phase, the system is said to be balanced. Under these circumstances, the current in the neutral is zero.

The power per phase is P

$$P = U_P \times I_P$$

The total power is the sum of the power in each phase.
The total power in a balanced circuit can be calculated:

$$P = \sqrt{3} U_L I_L$$

EXAMPLE A three-phase balanced load is connected in star, each phase of the load has an impedance of 10 Ω. The supply is 400 V 50 Hz.
Calculate the current in each phase (I_P):

$$I_P = \frac{U_P}{Z \times \sqrt{3}}$$

$$= \frac{400}{10 \times 1.732}$$

$$= 23 \text{ amperes per phase (in star } I_P = I_L).$$

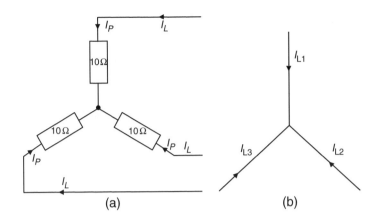

(a)   (b)

Fig. 66

Calculate the total power in one phase:

$$P = U_P \times I_P$$
$$= 231 \times 23$$
$$= 5313 \text{ watts}$$

Total power in all three phases $5313 \times 3 = 15\,939$ watts.
Total power in all three phases can also be calculated.

$$P = U_L \times I_L \times \sqrt{3}$$
$$= 400 \times 23 \times 1.732$$
$$= 15\,939 \text{ watts } (15.9\,\text{kW}).$$

DELTA-CONNECTED MOTORS (MESH)

For delta-connected loads (Figure 67) the voltage across the load will be the line voltage U_L, and the line current, I_L, will be the phase current I_P times $\sqrt{3}$.

As a calculation

$$I_L = I_P \times \sqrt{3}$$

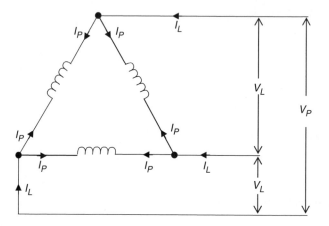

Fig. 67

The total power under these conditions is $P = \sqrt{3}U_L I_L$.

EXAMPLE (Using the same values as were used for star connection)

A three-phase balanced load is connected in delta, each phase of the load has an impedance of 10 Ω and the supply is 400 V 50 Hz.

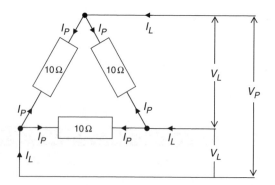

Fig. 68

Calculate the phase current and the load current:

$$I_P = \frac{U_L}{Z}$$
$$= \frac{400}{10}$$
$$I_P = 40\,\text{A}$$
$$I_L = I_P \times \sqrt{3}$$
$$= 40 \times 1.732$$
$$= 69.28\,\text{A}$$

The total power can be calculated

$$P = \sqrt{3} \times U_L \times I_L$$
$$= 1.732 \times 400 \times 69.28$$
$$= 47\,997\,\text{watts}\,(47.8\,\text{kW}).$$

It can be seen that the power dissipated in a delta-connected load is three times that of the star-connected load. The same applies to the current drawn from the supply.

RESISTANCE AND INDUCTANCE IN THREE-PHASE CIRCUITS

In many three-phase loads such as motors, inductance as well as resistance will need to be taken into account.

EXAMPLE 1 Three coils are connected in star formation to a 400 volt 50 Hz supply, each coil has a resistance of 35 Ω and an inductance of 0.07 H.
Calculate (a) the line current I_L and (b) the total power dissipated.
Step 1
Calculate inductive reactance

$$X_L = 2\pi f L$$
$$= 2 \times 3.142 \times 50 \times 0.07$$
$$= 22\,\Omega$$

An impedance triangle could be drawn if required as follows:

Draw to scale a line representing resistance on the horizontal (opposite), at right angles to line R. Draw a line representing inductive reactance (adjacent). The length of the hypotenuse will represent the impedance Z.

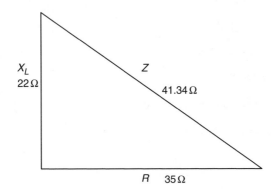

Fig. 69

By calculation

$$Z^2 = X_L^2 + R^2$$
$$Z^2 = 22^2 + 35^2$$
$$Z = \sqrt{22^2 + 35^2}$$
$$= 41.34 \ \Omega$$

Impedance (Z) is 41.34 Ω per phase.

Because the circuit has inductive reactance and resistance, the load will have a power factor, this must now be calculated:

$$\text{Power factor} = \frac{R}{Z}$$
$$= \frac{35}{41.34}$$
$$= 0.84$$

Line current can now be calculated

$$I_L = \frac{U_L}{Z \times \sqrt{3}} \; (U_L \text{ is line voltage})$$

$$= \frac{400}{41.34 \times 1.732 \times 0.84}$$

$$= 5.59 \, \text{A} \; (\text{in star } I_L = I_P. \text{ Remember this is the} \\ \text{current per phase})$$

Power can now be calculated

$$P = \sqrt{3} \times U_L \times I_L \times \cos\phi$$

$$= 1.732 \times 400 \times 5.59 \times 0.84$$

$$= 3250 \, \text{watts} \; (3.25 \, \text{kW})$$

EXAMPLE 2   Using the coils in the previous example connected in delta, calculate the line current and total power dissipated.
In delta

$$I_P = \frac{U_L}{Z \times \cos\phi}$$

$$= \frac{400}{41.34}$$

$$I_P = 9.68 \, \text{A}$$

$$I_L = I_P \times \sqrt{3}$$

$$= 9.68 \times 1.732$$

$$I_L = 16.76 \, \text{A} \; (\text{Note: three times the current in as in-star})$$

Power can now be calculated

$$P = \sqrt{3} \times U_L \times I_L \times pf$$

$$= 1.732 \times 400 \times 16.76 \times 0.84$$

$$= 9750 \, \text{watts} \; (9.75 \, \text{kW}) \; (\text{Note: three times the} \\ \text{power as in-star})$$

1. Three equal coils of inductive reactance $30\,\Omega$ and resistance $40\,\Omega$ are connected in-star to a three-phase supply with a line voltage of $400\,V$. Calculate (a) the line current and (b) the total power.

2. The load connected between each line and neutral of a $400\,V$ $50\,Hz$ three-phase circuit consists of a capacitor of $31.8\,\mu F$ in series with a resistor of $100\,\Omega$. Calculate (a) the current in each line and (b) the total power.

3. The load connected between each line and the neutral of a $400\,V$ three-phase supply consists of:

   (a) Between L1 and N, a noninductive resistance of $25\,\Omega$

   (b) Between L2 and N, an inductive reactance of $12\,\Omega$ in series with a resistance of $5\,\Omega$

   (c) Between L3 and N, a capacitive reactance of $17.3\,\Omega$ in series with a resistance of $10\,\Omega$.

   Calculate the current in each phase.

4. Three resistors each of $30\,\Omega$ are connected in-star to a $400\,V$ three-phase supply. Connected in-star to the same supply are three capacitors each with a reactance of $40\,\Omega$. Calculate (a) the resultant current in each line and (b) the total power.

5. Three capacitors, each with a reactance of $10\,\Omega$ are to be connected to a three-phase $400\,V$ supply for power factor improvement. Calculate the current in each line if they are connected (a) in-star, (b) in-mesh.

6. A $440\,V$, three-phase, four-wire system supplies a balanced load of $10\,kW$. Three single-phase resistive loads are added between lines and neutrals as follows: (a) L1–N $2\,kW$, (b) L2–N $4\,kW$, (c) L3–N $3\,kW$. Calculate the current in each line.

7. Three $30\,\Omega$ resistors are connected (a) in-star, (b) in-delta to a $400\,V$ three-phase system. Calculate the current in each resistor, the line currents and the total power for each connection.

8. Each branch of a mesh connected load consists of a resistance of 20 Ω in series with an inductive reactance of 30 Ω. The line voltage is 400 V. Calculate (a) the line currents and (b) total power.

9. Three coils each with a resistance of 45 Ω and an inductance of 0.2 H, are connected to a 400 V three-phase supply at 50 Hz, (a) in-mesh (b) in-star. Calculate (i) the current in each coil and (ii) the total power in the circuit.

10. A three-phase load consists of three similar inductive coils, each with a resistance of 50 Ω and an inductance of 0.3 H. The supply is 400 V 50 Hz. Calculate (i) the current in each line, (ii) the power factor, (iii) the total power, when the load is (a) star-connected, (b) delta-connected.

11. Three equal resistors are required to absorb a total of 24 kW from a 400 V three-phase system. Calculate the value of each resistor when they are connected (a) in-star, (b) in-mesh.

12. To improve the power factor, a certain installation requires a total of 48 kVAr equally distributed over the three phases of a 415 V 50 Hz system. Calculate the values of the capacitors required (microfarads) when the capacitors are connected (a) in-star (b) in-delta.

13. The following loads are connected to a three-phase 400 V 50 Hz supply. A noninductive resistance of 60 Ω is connected between L1 and L2, an inductive reactance of 30 Ω is connected between L2 and L3, and a capacitor of 100 μF is connected between L1 and L3. Calculate the total power and the current through each load.

14. A motor generator set consists of a d.c. generator driven by a three-phase a.c. motor. The generator is 65% efficient and delivers 18 A at 220 V. The motor is 75% efficient and operates at 0.5 p.f. lagging from a 415 V supply. Calculate (a) the power output of the driving motor, (b) the line current taken by the motor.

15. A conveyor raises 1600 kg of goods through a vertical distance of 5 m in 20 s. It is driven by a gear which is

55% efficient. Calculate the power output of the motor required for this work. If a three-phase 400 V motor with an efficiency of 78% is fitted, calculate the line current assuming a power factor of 0.7.

16. A 415 V three-phase star-connected alternator supplies a delta-connected induction motor of full load efficiency 87% and a power factor of 0.8. The motor delivers 14 920 W. Calculate (a) the current in each motor winding, (b) the current in each alternator winding, (c) the power developed by the engine driving the alternator, assuming that the alternator is 82% efficient.

17. A three-phase transformer supplies a block of flats at 230 V line to neutral. The load is balanced and totals 285 kW at 0.95 power factor. The turns ratio of the transformer, primary to secondary is 44:1 and the primary side of the transformer is connected in mesh. Calculate the primary line voltage. Draw a diagram and mark the values of the phase and line currents in both windings.

THREE-PHASE CIRCUITS

In a balanced three-phase circuit no current will flow in the neutral. In an unbalanced three-phase circuit, some current will flow in the neutral, this current can be calculated by using four different methods.

EXAMPLE 1 A sub-main supplying an unbalanced three-phase and neutral distribution board has currents of 75 A in L1 (brown), 55 A in L2 (black) and 40 A in L3 (grey). Calculate the current in the neutral.

Draw L1, L2 and L3 to scale at 120° to each other.

Now draw a vertical line from the end of L2 the same length as L1. Then join the tops of L1 and L2.

Draw a line between the shortest angles of the parallelogram. Now draw a line the same length and parallel with L3 from the top left angle. Join the ends of L3 and the line parallel to it. Measure the

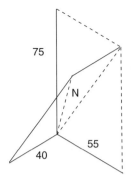

Fig. 70

gap between the shortest angle and, if the phasor is drawn to scale, this will be the current flowing in the neutral.

EXAMPLE 2   Using a triangle, draw a horizontal line to scale to represent L1.

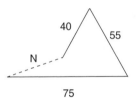

Fig. 71

From the right-hand side of this line, draw a line at 120° to it to scale to represent L2. Now draw a line from the end of L2 at 120° to it to represent L3 to scale.

Measure the gap between the open ends of L1 and L3, this will be the current flowing in the neutral.

EXAMPLE 3   Using a simpler phasor diagram and a simple calculation.

Take the smallest current from the other two currents.

In this example L3 (40 A) is the smallest current.

(L1) $75\,\text{A} - 40\,\text{A} = 35\,\text{A}$

(L2) $55\,\text{A} - 40\,\text{A} = 15\,\text{A}$

Draw a vertical line to scale to represent corrected L1.

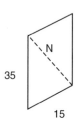

35

15

Fig. 72

Now draw a line to scale at 120° to it to represent corrected L2.

Draw a line vertically from the end of L2 the same length as L1, join open ends.

Measure between the shortest angles and this will be the current in the neutral.

EXAMPLE 4 By calculation only

Subtract the smallest current from the other two.

(L1) $75\,\text{A} - (\text{L3})\ 40\,\text{A} = 35\,\text{A}$

(L2) $55\,\text{A} - (\text{L3})\ 40\,\text{A} = 15\,\text{A}$

Current in the neutral can now be calculated (this will not be exact but very close):

$$\sqrt{35^2 - 15^2} = 31.62$$

EXERCISE 11

1. Three separate single-phase loads are to be connected to a three-phase and neutral supply, the currents in the loads are as follows: L1 = 32A, L2 = 24 A and L3 = 30A. Calculate the current flowing in the neutral.

2. A submain is to be installed to supply the following loads. L1 is a lighting load of 3.2 kW, L2 is a cooker load of 7 kW, L3 is supplying a 20 A power load. The supply voltage is 230 V 50 Hz.

3. Calculate the current flowing in the neutral conductor of a supply cable when the following currents are flowing in the phase conductors: L1 = 10 A, L2 = 30 A, L3 = 20 A.

THREE-PHASE POWER

The chapter entitled Power factor in Book 1 explains leading and lagging power factor. In this chapter, we will take power factor a step further and see how it affects three-phase circuits.
It is often simpler to draw a power triangle showing:

Active power or true power in watts (W) or kW.

Apparent power in volt amps (VA) or kVA.
Reactive power in VAr or kVAr.

Power factor cos ϕ is found

$$\frac{kW}{kVAr}$$

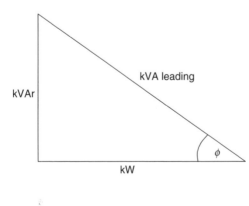

Fig. 73

Leading power factor

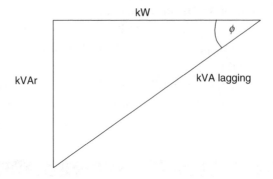

Fig. 74

Lagging power factor

To find the true power, apparent power and the reactive power we must first calculate as follows:

EXAMPLE I   A three-phase motor with an output of 2.8 kW and a power factor of 0.89 (p.f.) is connected to a 400 V 50 Hz supply. Calculate:

(a) the power drawn from the supply

(b) the reactive power

(c) the line current

(a) $\text{p.f.} = \dfrac{\text{kW}}{\text{kVA}}$

Transposed

$$\text{kVA} = \frac{\text{kW}}{\text{p.f.}}$$

$$= \frac{2.8}{0.89}$$

$= 3.14\,\text{kVA}$ (power drawn from supply is greater than the power delivered by the motor)

(b) $\text{kVAr}^2 = \text{kVA}^2 - \text{kW}^2$ or

$\text{kVAr} = \sqrt{\text{kVA}^2 - \text{kW}^2}$

$= \sqrt{3.14^2 - 2.8^2}$

$= 1.42\,\text{kVAr}$

Enter into calculator

$$3.14X^2 - 2.8X^2 = \sqrt{} = (\text{answer})$$

(c) Line current calculation

$$P = \sqrt{3}U_L I_L \cos\phi$$

$$= \frac{W}{\sqrt{3} \times U_L \times \cos\phi}$$

Transposed

$$= \frac{2800}{\sqrt{3} \times 400 \times 0.89}$$

$$= 4.54\,\text{A}$$

EXAMPLE 2   A commercial building is supplied by a three-phase, four-wire 400 V 50 Hz supply and the phases are loaded as follows:

L1 is taking 35 kW at unity power factor

L2 is taking 40 kVA at 0.8 lagging power factor

L3 is taking 60 kVA at 0.7 leading power factor

Calculate the power factor for the system.

Phase L1 is unity power factor

$$\text{p.f.} = \frac{\text{kW true power}}{\text{kVA apparent power}}$$

Transposed

$$\text{kVA} = \frac{\text{kW}}{\text{p.f.}}$$

$$= \frac{35}{1}$$

$$= 35\,\text{kVA}$$

This phase has no reactive power.

The circuit is purely resistive and can be shown by phasor as

Fig. 75

Phase L2 has a power factor of 0.8 lagging.
Using formulae

$$\text{p.f.} = \frac{\text{kW}}{\text{kVA}}$$

Transposed to find kW

$$= \text{p.f.} \times \text{kVA}$$

$$= 0.8 \times 40$$

$$= 32\,\text{kW}$$

Reactive (kVAr) power can be calculated

$$\text{kVAr} = \sqrt{\text{kVA}^2 - \text{kW}^2}$$

$$= \sqrt{40^2 - 32^2}$$

$$= 24\,\text{kVAr}.$$

The circuit is inductive (lagging) and can be shown by phasor as:

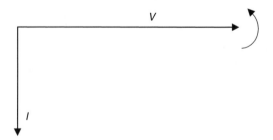

Fig. 76

Phase L3 has a power factor of 0.7 leading.

Using the same formulae reactive power for phase L3

$$\text{p.f.} \times \text{kW} = 0.7 \times 60$$

$$= 42 \, \text{kW}$$

$$\text{kVAr} = \sqrt{60^2 - 42^2}$$

$$= 42.84 \, \text{kVAr leading.}$$

This capacitive circuit can now be shown as a phasor.

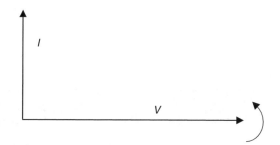

Fig. 77

This circuit is capacitive and is a leading circuit. We must now decide if the system is lagging or leading. As the lagging component of the system is greater than the leading, it will be a lagging system. The total kVAr will be the difference between the leading and lagging component.

Lagging 24 kVAr

Leading 42.84 kVAr

$42.84 - 24 = 18.84 \, \text{kVAr.}$

Total power is calculated by adding the power L1 + L2 + L3

L1 $= 35 \, \text{kW}$

L2 $= 32 \, \text{kW}$ $\quad (40 \, \text{A} \times 0.8)$

L3 $= 42 \, \text{kW}$ $\quad (60 \, \text{A} \times 0.7)$

Total power 109 kW.

Total reactive component 18.84 kVAr (leading).

kVA can now be calculated

$$\sqrt{18.84^2 + 109^2}$$
$$= 110.60 \, \text{kVA}$$

Power factor can now be calculated

$$\text{p.f.} = \frac{\text{kW}}{\text{kVA}}$$
$$= \frac{109}{110.60}$$
$$= 0.98$$

VOLTAGE DROP IN THREE-PHASE CIRCUITS

The method explained here is not a rigid treatment of three-phase voltage drop, but will provide a result which will be sufficient for most purposes (reactance is not taken into account).

BS 7671 allows a maximum voltage drop in an installation of 4% for each circuit. This is to be measured from the supply intake to the furthest point of the circuit. In most installations the line voltage (U_L) of a three-phase supply is 400 volts.

4% of 400 = 16 volts.

If the three-phase circuit is not a balanced load, each phase should be calculated as a single-phase circuit

4% of 230 = 9.2 volts

EXAMPLE 1   A three-phase balanced load of 15 A per phase is supplied by a steel wire armoured cable with a c.s.a. of 2.5 mm^2. The volt drop for this cable is 15 mV/A/m and the circuit is 30 m long.

Calculate (a) the voltage drop in the cable and (b) determine if it complies with the voltage drop requirements of BS 7671.

(a) $\text{Volt drop} = \dfrac{\text{mV} \times I \times L}{1000}$

$= \dfrac{15 \times 15 \times 40}{1000}$

$= 9 \text{ V per phase}$

The equivalent voltage drop in line voltage is

$= \sqrt{3} \times 9$

$= 15.58 \text{ volts}$

(b) This complies as it is less than 16 V.

EXAMPLE 2   A three-phase 10 kW motor operates on full load with efficiency 80% and power factor 0.75. It is supplied from a switchboard through a cable each core of which has resistance 0.2 Ω. Calculate the voltage necessary at the supply end in order that the voltage at the load end terminals shall be 400 V.

The full load current of the motor is

$$I = \dfrac{10 \times 1000}{\sqrt{3} \times 400 \times 0.75} \times \dfrac{100}{80}$$

$\underline{= 24.06 \text{ A}}$

The voltage drop per core of the cable

$= 24.06 \times 0.2$

$= 4.8 \text{ V}$

The equivalent reduction in the line voltage

$= \sqrt{3} \times 4.8$

$= 8.31 \text{ V}$

The required the voltage at the switchboard

$= 400 + 8.31$

$\underline{= 408.3 \text{ V}}$

This is not a rigid treatment of the problem but the method gives a result sufficiently accurate for most practical purposes.

EXAMPLE 3   The estimated load in a factory extension is 50 kW balanced at 0.8 p.f. The supply point is 120 metres away and the supply voltage is 400 V. Calculate the cross-sectional area of the cable in order that the total voltage drop shall not exceed 2.5% of the supply voltage.

Take the resistivity of copper as 1.78×10^{-8} Ωm.
The line current

$$I = \frac{50 \times 1000}{\sqrt{3} \times 400 \times 0.8}$$

$$= 90.21\,\text{A}$$

Allowable reduction in line voltage

$$= 2.5\% \times 400$$

$$= 10\,\text{V}$$

Equivalent reduction in phase voltage

$$= \frac{10}{\sqrt{3}}$$

$$= 5.77\,\text{V}$$

Resistance per core of the cable

$$= \frac{5.77}{90.21}$$

$$= 0.06\,396\,\Omega$$

The resistance of a cable is given by

$$R = \frac{\rho l}{A}$$

where ρ is the resistivity (Ωm)
           l is the length (m)
           A is the cross-sectional area (m^2)

so that

$$A = \frac{\rho l}{R}$$

$$A = \frac{1.78\,\Omega\text{m} \times 120\,\text{m}}{10^8 \times 0.06396\,\Omega}$$

$$= \frac{0.33}{10^4}\,\text{m}^2$$

$$= \frac{0.33}{10^4}\,\text{m}^2 \left[\frac{10^6\text{mm}^2}{1\text{m}^2}\right]$$

$$= \underline{33\,\text{mm}^2}$$

EXAMPLE 4 A p.v.c. trunking is to be used to enclose single-core p.v.c.-insulated distribution cables (copper conductors) for a distance of 30 m from the main switchgear of an office building to supply a new 400 V T.P. and N distribution fuseboard. The balanced load consists of 24 kW of discharge lighting. The fuses at the main switch-fuse and at the distribution board are to BS 88 part 2. The voltage drop in the cables must not exceed 6 V. The ambient temperature is anticipated to be 35 °C. The declared value of I_p is 20 kA and that of Z_e is 0.30 Ω. Assume that the requirements of BS 7671 434-03 are satisfied by the use of BS 88 fuses.

(a) For the distribution cables, establish the:

  (i) design current (I_b)

  (ii) minimum rating of fuse in the main switch-fuse (I_n)

  (iii) maximum mV/A/m value

  (iv) minimum current rating (I_t)

  (v) minimum cross-sectional area of the live conductors

  (vi) actual voltage drop in the cables.

(b) It is proposed to install a 2.5 mm^2 protective conductor within the p.v.c. trunking. Verify that this meets shock protection requirements. (C & G)

(a) (i) Design current $I_b = \dfrac{24 \times 10^3 \times 1.8}{\sqrt{3} \times 400}$ (1.8 factor for discharge lighting)

$$= 62.36\,\text{A}$$

(ii) Minimum BS 88 fuse rating (I_n) is 63 A.

(iii) Maximum mV/A/m value $= \dfrac{6 \times 1000}{62.36 \times 30}$

$$= 3.2\,\text{mV/A/m}$$

(iv) Minimum current rating $(I_t) = \dfrac{63}{0.94}$ (temperature correction factor C_a for 35 °C)

$$= 67.02\,\text{A}$$

(v) Minimum c.s.a. of cable is $16\,\text{mm}^2$ (68 A 2.4 mV/A/m).

(vi) Actual voltage drop in 30 m $= \dfrac{2.4 \times 62.36 \times 30}{1000} = 4.49\,\text{V}$

(b) Check compliance with Table 41D (BS 7671) using *IEE On-Site Guide.*

From Table 9A, $R_1 + R_2$ for $16\,\text{mm}^2/2.5\,\text{mm}^2 = 1.15 + 7.41\,\text{m}\Omega/\text{m}$.
From Table 9C, factor of 1.20 must be applied.
Now $Z_s = Z_e + R_1 + R_2$

$$R_1 + R_2 = \dfrac{30 \times (1.15 + 7.41) \times 1.20}{1000}$$

$$= 0.308\,\Omega$$

∴ $\qquad Z_s = 0.3 + 0.308$

$$Z_s = 0.608\,\Omega$$

This satisfies Table 41D as the maximum Z_s for a 63 A fuse is 0.86 Ω.

EXAMPLE 5 It is proposed to install a new 230 V 50 Hz distribution board in a factory kitchen some 40 m distant from the supplier's intake position.

It is to be supplied by two $25\,\text{mm}^2$ p.v.c. insulated (copper conductors) single-core cables in steel conduit. Protection at origin of the cables is to be by BS 88 fuses rated at 80 A.

It is necessary for contractual purposes to establish:

(a) the prospective short circuit current (p.s.c.c.) at the distribution board, and

(b) that the proposed distribution cables will comply with BS 7671 requirements 434-03-03.

A test conducted at the intake position between phase and neutral to determine the external impedance of the supplier's system indicates a value of 0.12 Ω.

(a) The resistance of distribution cables from intake to distribution board

From Table 9A (*IEE On-Site Guide*), R_1/R_2 for $25\,\text{mm}^2/25\,\text{mm}^2$ cables $= 1.454\,\text{mV/m}$.

From Table 9C a multiplier of 1.20 is necessary using the Table 6A figures as

$$R_1/R_2 = \frac{40 \times 1.454 \times 1.20}{1000}$$

$$= 0.0698\,\Omega \text{ (regard this as impedance)}$$

So total short circuit fault impedance $= 0.12 + 0.0698$
$$= 0.19\,\Omega$$

Thus $\quad I_f = \dfrac{230}{0.19}$

∴ p.s.c.c. $= 1210\,\text{A}$

From Appendix 3, Figure 3.3A, the BS 88 fuse clearance time is approximately 0.1 s.

(b)

From Requirement 434-03-03, $t = \dfrac{k^2 S^2}{I^2}$

$$= \frac{115^2 \times 25^2}{1200^2}$$

∴ limiting time for conductors $(t) = 5.74\,\text{s}$

The cables are disconnected well before the $25\,\text{mm}^2$ cable conductors reach their limiting temperature, thus they are protected thermally.

EXAMPLE 6 Two $25\,\text{mm}^2$ single-core p.v.c.-insulated cables (copper conductors) are drawn into p.v.c. conduit along with a $10\,\text{mm}^2$ protective conductor to feed a 230 V industrial heater.

The following details are relevant:

Protection at the origin is by 80 A BS 88 fuses.

The tested value of Z_e at the cables origin is $0.35\,\Omega$.

The length of cables run is 55 m.

(a) Establish the:
  (i) value of $R_1 + R_2$ of the cables
  (ii) prospective earth fault loop current (I_{ef})
  (iii) the disconnection time of the fuse.
(b) Does the clearance time comply with BS 7671?

(a) (i) Using the *IEE On-Site Guide*

From Table 9A $R_1 + R_2$ for $25\,\text{mm}^2/10\,\text{mm}^2$ cables $= 2.557\,\text{m}\Omega/\text{m}$. From Table 9C apply the factor 1.20

$$\text{Thus } R_1 + R_2 = \frac{55 \times 2.557 \times 1.20}{1000}$$

$$= 0.169\,\Omega$$

$$\text{So } Z_s \text{ at distribution board} = 0.35 + 0.169$$

$$= 0.519\,\Omega$$

(ii) Prospective earth fault current $(I_{ef}) = \dfrac{230}{0.519}$

$$= 443\,\text{A}$$

(iii) Using BS 7671, Appendix 3, Table 3.3 A, disconnection time is 3.8 s.

(b) The clearance time complies with BS 7671 Requirement 413-02-13 which specifies a maximum disconnection time of 5 s.

EXAMPLE 7 It is necessary to confirm that the cross-sectional area of the protective conductor in a previously installed 400/230 V distribution circuit complies with BS 7671 requirement 543-01-03. The phase conductors are $10\,\text{mm}^2$ and the circuit-protective conductor is $2.5\,\text{mm}^2$. The length of the cables run in plastic conduit is 85 m. Protection by 32 A, BS 88 fuses and the value of Z_e is 0.4 Ω.

Using the *IEE On-Site Guide*

From Table 9A $R_1 + R_2$ for $10\,\text{mm}^2/2.5\,\text{mm}^2$ cables $= 1.83 + 7.41\,\text{m}\Omega/\text{m}$

From Table 9C apply the factor 1.20

$$\text{Thus} \quad R_1 + R_2 = \frac{85 \times (1.83 + 7.41) \times 1.20}{1000}$$

$$= 0.942\,\Omega$$

So Z_s at distribution board $= 0.4 + 0.942$

$$= 1.342\,\Omega$$

Prospective earth fault loop current $(I_{ef}) = \dfrac{230}{1.342}$

$$= 171\,\text{A}$$

Using BS 7671

From Appendix 3 Table 3.3A, the fuse clearance time is 0.9 s.

From BS 7671 requirement 543-01-03

$$s = \frac{\sqrt{I^2 t}}{k} \quad (k \text{ is 115 Table 54C})$$

$$= \frac{\sqrt{171^2 \times 0.9}}{115}$$

$$= \underline{1.41\,\text{mm}^2}$$

This confirms that a $2.5\,\text{mm}^2$ protective conductor is acceptable.

EXAMPLE 8   The declared value of I_p at the origin of a 230 V 50 Hz installation is 1.5 kA. The length of $25\,\text{mm}^2$ p.v.c./p.v.c. metre tails is 2 m; at this point a switch-fuse containing 100 A BS 88 Part 2 fuses is to be installed to provide control and protection for a new installation. A 20 m length of $16\,\text{mm}^2$ heavy duty mineral insulated cable (exposed to touch), (copper conductors and sheath) is to be run from the switch-fuse to a new distribution board.

(a) Establish that the mineral cable complies with BS 7671 requirement 434-03.

(b) How could you ensure that the requirements of BS 7671 413-02-01, etc. and Chapter 7 are satisfied?

(a) Resistance of 2 m of 25 mm^2 metre tails using mV/A/m value from Table 9D1A as ohms per metre at 70 °C

$$R_{mt} = 2 \times 0.00175$$
$$= 0.0035 \, \Omega$$

Resistance of 20 m of 16 mm^2 twin m.i.c.c. cable using mV/A/m values from Table 4JB as ohms per metre at 70 °C

$$R_{mi} = 20 \times 0.0026$$
$$= 0.052 \, \Omega$$

$$\text{Impedance of supply} = \frac{230}{1500}$$
$$= 0.153 \, \Omega$$

Thus total impedance from source to distribution board

$$= 0.153 + 0.0035 + 0.052 = 0.2085 \, \Omega$$

Prospective short circuit fault current

$$I_p = \frac{230}{0.2085} = 1103 \, \text{A}$$

Disconnection time from BS 7671 Table 3.3B is 0.3 s
Now using the 434-03-03 adiabatic equation

$$t = \frac{k^2 S^2}{I^2}$$
$$= \frac{135^2 \times 16^2}{1103^2}$$
$$= 3.8 \, \text{s}$$

Thus 16 mm^2 cable is protected against the thermal deterioration.

(b) As no details are available in BS 7671 in relation to the resistance/impedance of the m.i.c.c. sheath, the prospective value of Z_s could not be established, but the actual value must be tested when the installation is commissioned and the value recorded in the Electrical Installation Certificate referred to in requirement 742.

EXAMPLE 9   A 230 V, 50 Hz, 5 kW electric motor is fed from a distribution board containing BS 88 Part 2 fuses. The wiring between the d.f.b. and the motor starter which is 20 m distant is p.v.c.-insulated single-core cables drawn into steel conduit. Assume that the:

(i)   starter affords overload protection;
(ii)   motor has a power factor of 0.75 and an efficiency of 80%;
(iii) ambient temperature is 40 °C;
(iv)   fuse in the d.f.b. may have a rating up to twice the rating of the circuit cables;
(v)   volts drop in the motor circuit cables must not exceed 6 V;
(vi)   resistance of metal conduit is 0.1 Ω per metre;
(vii) 'worst' conduit run is 8 m with two 90° bends;
(viii) I_p at d.f.b. is 2 kA;
(ix)   value of Z_e is 0.19 Ω.

Establish the:

(a) design current (I_b);
(b) rating of circuit fuse;
(c) minimum cable rating (I_n) between d.f.b. and starter;
(d) minimum cable cross-sectional area;
(e) actual voltage drop in cables;
(f) prospective short circuit current;
(g) short circuit disconnection time;
(h) whether BS 7671 requirement 434-03-03, etc. is satisfied;
(i) whether BS 7671 requirement 413-02-04, etc. is satisfied;
(j) minimum conduit size.

(a) Design current $(I_b) = \dfrac{5000}{230 \times 0.75 \times 0.8}$

$$= 36.2\,\text{A}$$

(b) Rating of circuit fuse may be 80 A.

(c) Minimum cable rating may be 40 A.

(d) Minimum cable c.s.a. $= \dfrac{40}{0.87} = 46\,\text{A}$

from Table 4D1A select $10\,\text{mm}^2$ cable (57 A).

(e) Actual voltage drop:

from Table 4D1B mV/A/m value for $10\,\text{mm}^2$ is 4.4 thus

$$\text{volts drop} = \frac{36.2 \times 20 \times 4.4}{1000}$$

$$= 3.19\,\text{V}$$

(f) Impedance of supply cables to d.f.b.

$$= \frac{230}{2000} = 0.115\,\Omega$$

Using BS 7671 Tables 9A and 9C, resistance of circuit cables

$$= \frac{20 \times 3.66 \times 1.2}{1000}$$

$$= 0.09\,\Omega$$

thus total circuit impedance $= 0.115 + 0.09 = 0.205\,\Omega$

$$\text{Prospective short circuit current} = \frac{230}{0.205}$$

$$= 1122\,\text{A}$$

(g) Disconnection time from Figure 3.3A is 0.1 s

(h) Cable thermal capacity $t = \dfrac{k^2 \times S^2}{I^2}$

$$= \frac{115^2 \times 10^2}{1122^2}$$

$$= 1.05\,\text{s}$$

Thus $10\,\text{mm}^2$ cables are thermally safe.

(i) Now $Z_s = Z_e + R_1 + R_2$

Using BS 7671 Tables 9A and 9C,

$$\text{Resistance of } R_1 = \frac{20 \times 1.83}{1000} = 0.0366\,\Omega$$

Resistance of conduit $R_2 = 20 \times 0.01 = 0.2\,\Omega$

Thus $Z_s = 0.19 + 0.0366 + 0.2 = 0.4266\,\Omega$

$$I_{ef} = \frac{230}{0.4266} = 539\,A$$

From BS 7671, Figure 3.3A, disconnection time is 1.4 s; this being less than 5 s, protection is satisfactory.

(j) From Table 5C,

cable factor for $2 \times 10\,\text{mm}^2$ cables $= 2 \times 105 = 210$

From Table 5D select 25 mm conduit (factor 292).

EXERCISE 12

1. A balanced load of 30 A is supplied through a cable each core of which has resistance $0.28\,\Omega$. The line voltage at the supply end is 400 V. Calculate the voltage at the load end, the percentage total voltage drop and the power wasted in the cable.

2. Each core of a three-core cable, 164 m long, has a cross-sectional area of $35\,\text{mm}^2$. The cable supplies power to a 30 kW, 400 V, three-phase motor working at full load with 87% efficiency and power factor 0.72 lagging. Calculate:

   (a) the voltage required at the supply end of the cable;
   (b) the power loss in the cable.

   The resistivity of copper may be taken as $1.78 \times 10^{-8}\,\Omega\text{m}$ and the reactance of the cable may be neglected.

3. A 40 kW, 400 V, three-phase motor, running at full load, has efficiency 86% and power factor 0.75 lagging. The three-core cable connecting the motor to the switchboard is 110 m long and its conductors are of copper $25\,\text{mm}^2$ in cross-section.

   Calculate the total voltage drop in the cable, neglecting reactance.

If the cable runs underground for most of its length, choose a suitable type of cable for the purpose and give a descriptive sketch of the system of laying it.

The resistivity of copper may be taken as 1.78×10^{-8} Ωm.

4. The estimated load in a factory extension is 200 kW at 0.85 p.f. (balanced). The supply point is 75 m away where the line voltage is 400 V. Choose the most suitable size of cable from those given below in order that the total voltage drop shall not exceed 2.5% of supply voltage.

Cross-sectional areas

of available conductors (mm^2)   35   50   70   95
(Resistivity of conductor is 1.78×10^{-8} Ωm.)

5. A motor taking 200 kW at 0.76 p.f. is supplied at 400 V three-phase by means of a three-core copper cable 200 m long.

   (a) Calculate the minimum cable cross-sectional area if the voltage drop is not to exceed 5 V.

   (b) If the cable size calculated is non-standard, select from the table a suitable standard cable and calculate the actual voltage drop using that cable.

Standard cross-sectional areas of cable conductors (mm^2)
      300   400   500   630
(Resistivity of copper 1.78×10^{-8} Ωm.)

6. A three-phase current of 35 A is supplied to a point 75 m away by a cable which produces a voltage drop of 2.2 mV per ampere per metre. Calculate the total voltage drop.

The following question should be answered by reference to be appropriate tables in BS 7671 and/or in the *IEE On-Site Guide* to BS 7671.

7. A balanced load of 85 A is required at a point 250 m distant from a 400 V supply position. Choose a suitable cable (clipped direct) from Tables 4E4A and 4E4B in order that the total voltage drop shall be within the BS 7671 specified limit (ambient temperature 30 °C).

8. A 25 kW, 400 V three-phase motor having full load efficiency and power factor 80% and 0.85 respectively is

supplied from a point 160 m away from the main switchboard. It is intended to employ a surface-run, multicore p.v.c.-insulated cable, non-armoured (copper conductors). The ambient temperature is 30 °C and BS 88 fuses are to be employed at the main switchboard. Select a cable to satisfy the BS 7671 requirements.

9. The total load on a factory sub-distribution board consists of:

   10 kW lighting balanced over three phases, unity power factor; 50 kW heating balanced over three phases, unity power factor and 30 kW motor load having an efficiency 80%, power factor 0.8.

   The line voltage is 400 V and the supply point is 130 m distant. Protection at the origin of the cable (clipped direct) is by BS 88 fuses. The ambient temperature is 30°C. Select a suitable cable from Tables 4D2A and 4D2B, in order that the voltage drop shall not exceed 3% of the supply voltage.

10. Calculate the additional load in amperes which could be supplied by the cable chosen for question 9 with the voltage drop remaining within the specified limits.

11. A 12 kW, 400 V three-phase industrial heater is to be wired using single-core p.v.c.-insulated cables (copper conductors) 30 m in length drawn into a steel conduit. The following details may be relevant to your calculation.

    Ambient temperature 40°C

    Protection by BS 3036 (semi-enclosed) fuses.

    Voltage drop in the cables must not exceed 10 V.

    The contract document calls for a 2.5 mm^2 conductor to be drawn into the conduit as a supplementary protective conductor.

    The worst section of the conduit run involves two right-angle bends in 7 m. Establish the:

    (a) design current (I_b);
    (b) minimum fuse rating (I_n);
    (c) maximum mV/A/m value;
    (d) minimum live cable rating (I_t);

(e) minimum live cable c.s.a.;

(f) actual voltage drop;

(g) minimum conduit size.

12. The external live conductor impedance and external earth fault loop impedance are tested at the intake of a 230 V single-phase installation and show values of 0.41 Ω and 0.28 Ω, respectively. A p.v.c. trunking runs from the intake position to a distribution board 40 m distant and contains $35\,mm^2$ live conductors and a $10\,mm^2$ protective conductor.

(a) Estimate the:

    (i) prospective short circuit current (p.s.c.c.) at the distribution board;

    (ii) p.s.c.c. clearance time of the 100 A BS 88 fuse at the origin of the cable;

    (iii) value of the earth fault loop impedance (Z_s) at the distribution board;

    (iv) prospective earth fault loop current;

    (v) earth fault clearance time of the BS 88 fuse at the origin of the cable.

(b) State the maximum permitted value of Z_s under these conditions.

13. A p.v.c. trunking containing single-core p.v.c.-insulated distribution cables (copper conductors) is to be run 30 m from the 400/230 V main switchgear of an office building to supply a new T.P. and N distribution fuseboard. The balanced load consists of 24 kW of discharge lighting. The fuses at the main switch-fuse and at the distribution board are to BS 88 part 2. The voltage drop in the distribution cables must not exceed 6 V and the ambient temperature is anticipated to be 35 °C. The declared value of I_p is 20 kA and that of Z_e is 0.3 Ω. Assume that the requirements of BS 7671 434-5 are satisfied.

(a) For the distribution cables, establish and state the

    (i) design current;

    (ii) minimum rating of fuse in the main switch fuse;

(iii) maximum mV/A/m value;

(iv) minimum current rating;

(v) minimum cross-sectional area of the live conductors;

(vi) actual voltage drop in the cable.

(b) It is proposed to install a $4\,mm^2$ protective conductor within the p.v.c. trunking.

(i) State the value of Z_s.

(ii) Verify that this meets BS 7671 shock protection requirements.

14. A security building is to be built at the entrance to a factory. This new building is to be provided with a 230 V single-phase supply and is to be situated 20 m from the main switchroom. A 30 m twin p.v.c.-insulated armoured underground cable (copper conductors) supplies the new building, which allows 5 m at each end for runs within the main switchroom and security buildings. The connected load in the security building comprises

one 3 kW convector heater

two 1 kW radiators

two 1.5 kW water heaters (instantaneous type)

one 6 kW cooker

six 13 A socket outlets (ring circuit)

a 2 kW lighting load.

Diversity may be applied (business premises).

Establish:

(a) the prospective maximum demand;

(b) minimum current rating of the switch-fuse in the switchroom at the origin of the underground cable.

(c) Determine

(i) the minimum current rating of p.v.c.-insulated twin armoured (copper conductor) underground cable, assuming an ambient temperature of 25 °C and protection to be by BS 88 part 2 devices;

(ii) the minimum size of cable, assuming the voltage drop is limited to 2 V;

(iii) the actual voltage drop in the cable.   (C & G)

15. It is proposed to install a p.v.c.-insulated armoured cable to feed a 25 kW, 400 V, three-phases 50 Hz resistive element type of furnace. The cable is to be surface run along a brick wall of a factory and has a total length of 95 m. The protection at the origin of the circuit is to be by BS 88 fuses. The cable armour may be relied upon as the circuit protective conductor. The ambient temperature in the factory will not exceed 35 °C and the voltage drop must not exceed 10 V.

Determine and state the:

(a) design current;

(b) fuse rating;

(c) minimum cable current rating;

(d) maximum mV/A/m value;

(e) minimum cross-sectional area of the live conductors;

(f) actual voltage drop in the cable. (C & G)

16. A pipeline pump is connected to a 400/230 V three-phase supply. It is wired in $1.5 \, mm^2$ p.v.c.-insulated cables drawn into 25 mm steel conduit running 30 m from a distribution board containing BS 88 part 2 fuses (10 A fuses protect the pump).

It is now necessary to add alongside the pump a 12 kW 400/230 V heater and it is proposed to draw the p.v.c.-insulated cables into the existing 25 mm pump circuit conduit and insert suitable fuses into the distribution board which has vacant ways.

The following assumptions may be made:

(i) an ambient temperature of 35 °C;

(ii) the maximum distance between draw in boxes is 9.5 m with two right-angle bends;

(iii) the maximum voltage drop in the heater circuit is 5 V;

(iv) a $2.5 \, mm^2$ protective conductor is installed in the steel conduit to satisfy a clause in the electrical specifications.

Determine for the heater circuit the:

(a) design current I_b;

(b) suitable fuse rating I_n;

(c) maximum mV/A/m value;

(d) minimum cable current rating I_t;

(e) minimum cable c.s.a.;

(f) actual voltage drop.

For the existing pump circuit establish whether the:

(g) pump circuit cable current rating is still adequate;

(h) 25 mm conduit is suitable for the additional cables.

Voltmeters and ammeters: changing the use and extending the range

VOLTMETERS

The voltmeter is a high-resistance instrument and its essential electrical features may be represented by the equivalent circuit of Figure 78.

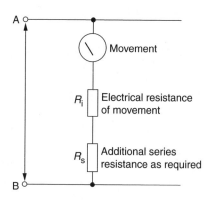

Fig. 78

R_i is the 'internal resistance' of the movement, i.e. the resistance of the moving coil or the resistance of the fixed coil in the case of a moving-iron instrument.

Independently of its resistance, the movement will require a certain current to deflect the pointer across the full extent of the scale against the effect of the controlling springs. This is the current required for full-scale deflection (f.s.d.).

The range of voltage which the instrument can indicate is governed by the total resistance R as measured between the terminals A and B, and

$$R = R_i + R_s$$

If $I_{f.s.d.}$ is the current required to produce full-scale deflection and R is the *total* resistance between A and B, the voltage between A and B at full-scale deflection is

$$U = R \times I_{f.s.d.}$$

$I_{f.s.d.}$ is fixed by the mechanical and electrical characteristics of the instrument and is not normally variable. The resistance R, however, can be fixed at any convenient value by adding the additional series resistance (R_s) as required.

EXAMPLE An instrument has internal resistance 20 Ω and gives f.s.d. with a current of 1 mA. Calculate the additional series resistance required to give f.s.d. at a voltage of 100 V.

$$U = R \times I_{f.s.d.}$$

$\therefore \quad 100 = R \times \dfrac{1}{1000}$ (note the conversion of milliamperes to amperes)

$\therefore \quad R = 100 \times 1000 \ \Omega$

$$= 100\,000 \ \Omega$$

But the instrument has internal resistance of 20 Ω; thus the additional resistance required is

$$R_s = R - R_i$$

$$= 100\,000 - 20$$

$$= 99\,980 \ \Omega$$

(Note that this is a somewhat unrealistic value in terms of what it is economically practical to manufacture. In practice, the additional resistance would be constructed to the nominal value of 100 000 Ω (100 kΩ) and slight adjustments would be made as necessary at the calibration stage to obtain f.s.d. with an applied 100 V.)

Any applied voltage less than 100 V of course produces a corresponding lower reading on the instrument.

The additional series resistor R_s is also known as a *multiplier*.

AMMETERS

The ammeter is a low-resistance instrument, and its equivalent electrical circuit is shown in Figure 79.

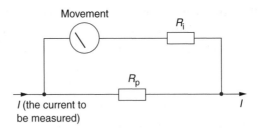

Fig. 79

The resistor R_p connected in parallel is the 'shunt' through which most of the current to be measured flows. Its value will be low compared with the internal resistance of the movement R_i. The calculation of R_p proceeds as follows.

Knowing R_i and the current required to give f.s.d., determine the voltage required to produce f.s.d. of the movement. For example, using the information of the previous example ($I_{f.s.d.} = 1$ mA and $R_i = 20$ Ω),

$$\text{p.d. required for f.s.d.} = U_{f.s.d.} = I_{f.s.d.} \times R_i$$

$$= \frac{1}{1000} \times 20$$

$$= \frac{20}{1000} \text{ V}$$

and this is the voltage drop which must be produced across the shunt resistor R_p.

The current which flows through R_p is the current to be measured minus the current which flows through the movement. If the greatest value of current to be measured is 20 A, the current which flows through R_p is $I = 20\,A - 1\,mA = 19.999\,A$. The voltage across R_p is then $V_{f.s.d.} = (20/1000)\,V$,

$$\therefore \quad R_p = \frac{U_{f.s.d.}}{I}$$

$$= \frac{20/1000}{19.999}$$

$$= 0.001\,\Omega$$

In fact, no significant difference is made if the 1 mA of total current which flows through the movement and not through the shunt is ignored in the calculation of R_p. Again, it is usual to manufacture the shunt to the nominal value calculated above and then to make slight adjustments at the calibration stage to obtain the desired full-scale deflection.

EXAMPLE   A moving-coil instrument gives full-scale deflection with a current of 1.2 mA, and its coil has resistance 40 Ω. Determine

(a) the value of the multiplier required to produce a voltmeter reading up to 50 V,

(b) the value of the shunt required to convert the instrument to an ammeter reading up to 10 A.

(a) Total resistance required to restrict the current to 1.2 mA from a 50 V supply is

$$R = \frac{50\,V}{(1.2/1000)\,A} \quad \text{(note the conversion to amperes)}$$

$$= 41\,667\,\Omega$$

The accurate value of additional series resistance required is

$$R_s = 41\,667 - 40$$

$$= 41\,627\,\Omega$$

(b) Voltage required to produce f.s.d. $= \dfrac{1.2}{1000} \times 40$

$$= \dfrac{4.8}{1000}\,\text{V}$$

Then $\quad \dfrac{4.8\,\text{V}}{1000} = R_{\text{p}} \times (10\,\text{A} - 0.0012\,\text{A})$

$\therefore \qquad R_{\text{p}} = \dfrac{4.8}{1000 \times 9.9988}$

$$= 4.8 \times 10^{-4}\,\Omega$$

Again, the 1.2 mA of current which flows through the instrument could have been neglected in calculating R_{p}.

EXERCISE 13

1. The coil of a moving-coil instrument has resistance 50 Ω, and a current of 0.8 mA is required to produce full-scale deflection. Calculate the voltage required to produce full-scale deflection.

2. A moving-coil instrument movement was tested without either shunt or multiplier fitted and it was found that at full-scale deflection the current through the coil was 1.15 mA and the voltage across it was 52 mV. Determine the resistance of the coil.

3. A moving-coil instrument gives full-scale deflection with a current of 1.5 mA and has resistance (without shunt or multiplier) of 25 Ω. Determine the value of additional series resistance (the multiplier) required to produce a voltmeter capable of measuring up to 150 V.

4. Using the instrument movement of question 3, modify the metre to measure currents up to 25 A by calculating the value of a suitable shunt resistor.

5. Given an instrument movement of resistance 40 Ω and requiring a current of 1 mA to produce f.s.d., determine the values of the various resistors required to produce a multi-range instrument having the following ranges:
   Voltage: 0–10 V, 0–150 V, 0–250 V
   Current: 0–1 A, 0–10 A

6. A moving-coil instrument requires 0.75 mA of current to produce f.s.d. at a voltage of 50 mV. The resistance of the coil is

(a) 0.015 Ω (b) 0.067 Ω (c) 66.7 Ω (d) 66 700 Ω

7. The coil of a moving-coil instrument has resistance 45 Ω and requires a current of 1.15 mA to produce f.s.d. The p.d. required to produce f.s.d. is

(a) 51.8 V   (b) 39 V   (c) 0.025 V   (d) 51.8 V

8. The value of the multiplier required to convert the instrument of question 7 to a voltmeter to measure up to 250 V is appxorimately

(a) 217 kΩ (b) 6.25 Ω (c) 217 Ω (d) 288 kΩ

9. The approximate value of the shunt required to convert the instrument of question 6 to an ammeter to measure up to 25 A is

(a) 2 Ω (b) 0.002 Ω (c) 0.067 Ω (d) 0.125 Ω

Alternating current motors

For a single phase motor:

Power = voltage × current × power factor

$$P = U \times I \times p.f.$$

For a three-phase motor:

$$\text{Power} = \sqrt{3}\,\frac{\text{line}}{\text{voltage}} \times \frac{\text{line}}{\text{current}} \times \frac{\text{power}}{\text{factor}}$$

$$P = \sqrt{3} \times U_L \times I_L \times p.f.$$

EXAMPLE 1   Calculate the current taken by a 1.7 kW 230 V single-phase motor working at full load with a power factor of 0.8 and an efficiency of 75%.

It should be noted that when a motor power rating is given, it is the output power unless otherwise stated.

Output is 1.7 kW or 1700 W.

Calculation for current drawn is

$$P = U \times I \times p.f.$$

$$\frac{P}{U \times p.f.}$$

$$\frac{\text{Output} \times 100}{U \times p.f. \times \text{efficiency}}$$

It is an easier calculation if the efficiency is shown as a decimal and put on the bottom line:

$$\frac{\text{Output}}{U \times p.f. \times \text{efficiency}}$$

$$= \frac{1700}{230 \times 0.8 \times 0.75}$$

Enter into calculator

$$1700 \div (230 \times 0.8 \times 0.75) = (\text{answer})$$

$$= 12.31 \, \text{A}.$$

EXAMPLE 2   A three-phase 400 V induction motor with an output of 12.4 kW is to be installed to drive a conveyor belt. The motor has a power factor of 0.85 and an efficiency of 78%. It is to be protected by a BS EN 60898 mcb type C. Calculate the current drawn per phase and the size of the protective device.

$$P = \sqrt{3}U_L I_L \times p.f.$$

$$I_L = \frac{P}{\sqrt{3} \times U_L \times p.f. \times \text{eff}}$$

$$= \frac{12\,400}{\sqrt{3} \times 400 \times 0.85 \times 0.78} \quad \begin{array}{l}\text{(Note 78\% changed to 0.78} \\ \text{and put on bottom)}\end{array}$$

Enter into calculator

$$12\,400 \div (\sqrt{3} \times 400 \times 0.85 \times 0.78) = (\text{answer})$$

(note the change of % to a decimal)

$$= 26.99 \, \text{A}.$$

Remember, the protective device must be equal to or greater than the design current (current drawn per phase). Table 41B2 in Chapter 41 of BS 7671 lists the sizes of protective devices.

In this case, a 32 A device should be used.

EXAMPLE 3 A three-phase 400 V induction motor connected in-star has an output of 18 kW and a power factor of 0.85. The motor circuit is to be protected by a BS 88 fuse. Calculate
(a) the design current (the current drawn from the supply), and
(b) the correct rating of the protective device for this circuit.

(a) Design current

$$I_L = \frac{P}{\sqrt{3} \times U_L \times p.f.}$$

$$= \frac{18\,000}{\sqrt{3} \times 400 \times 0.85}$$

$$I_L = 30.56\,\text{A}$$

(b) Protective device is 32A (if unsure of the ratings of fuses, Table 41D in BS 7671 can be used for fuses).

EXAMPLE 4 The same motor as in Example 3 is connected in delta. Calculate (a) design current, (b) the output power, and (c) the correct size of protected device.

Remember, the output of the motor will increase if it is connected in delta.

(a) Design current I_L in delta

$$30.56 \times 3 = 91.7\,\text{A} \ (3 \times \text{current in star})$$

(b) Output power

$$P = \sqrt{3} \times U_L \times I_L \times p.f.$$

$$= \sqrt{3} \times 400 \times 91.7 \times 0.85$$

$$\text{Output power} = 54\,000 \text{ watts or } 54\,\text{kW}$$

$$(3 \times \text{output in star})$$

(c) Protective device = 100 A.

Examples 3 and 4 show by calculation that more current will be drawn from the supply when connected in delta. This is the reason why it is common for three-phase motors to be started in star and then changed to delta.

The starting current for motors is considerably greater than their running current, between 5 and 10 times depending on the load being driven.

It can be seen that in delta, the start current for the motor will be at least:

$30 \times 5 = 150$ A when started in star.

$53 \times 5 = 265$ A when started in delta.

This is why particular care should be taken when selecting protective devices. It is important that a device which can handle medium to high in-rush currents is selected.

EXAMPLE 5 A 400 V three-phase motor with a power factor of 0.7 has an output of 3.2 kW. Calculate (a) the line current, (b) the power input of motor (kVA) and (c) the reactive component of motor kVAr.

$$P = \sqrt{3U_L}\, I_L\, p.f.$$

Transpose

(a) $\dfrac{3200}{\sqrt{3} \times 400 \times 0.7} = 6.6\,\text{A}$

(b) $p.f. = \dfrac{\text{output}}{\text{input}}$ or $\dfrac{\text{kW}}{\text{kVA}}$

Transpose for power factor

$$\text{kVA} = \dfrac{\text{kW}}{p.f.}$$

$$= \dfrac{3200}{0.7}$$

$$\text{kVA} = 4571\,\text{VA} \text{ or } 4.571\,\text{kVA}$$

(c) $$kVAr^2 = kVA^2 - kW^2$$

$$= \sqrt{kVA^2 - kW^2}$$

$$= \sqrt{4.571^2 - 3.2^2}$$

$$kVAr = 3.26$$

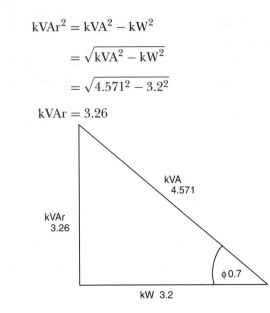

Fig. 80

EXAMPLE 6   A load of 300 kg is to be raised through a vertical distance of 12 m in 50 seconds by an electric hoist with an efficiency of 80%. Calculate the output required by a motor to perform this task.

The force required to lift a mass or load of 1 kg against the force of gravity is 9.81 N.

$$\text{Work on load} = \text{force} \times \text{distance}$$

$$12 \times 300 \times 9.81 = 35\,316 \text{ Newton metres}$$

$$1\,Nm = 1\,Joule.$$

Therefore, work required to be done to lift the load is 35 316 Joules.

$$\begin{aligned} \text{Output required} \\ \text{by motor} \end{aligned} = \frac{\text{energy in joules}}{\text{time in seconds}}$$

$$= \frac{35\,316}{50 \times 0.8} \quad \begin{array}{l}\text{(efficiency of hoist used as a}\\ \text{decimal, always under to}\\ \text{increase output)}\end{array}$$

$$= 882.9 \text{ watts or joules per second.}$$

EXAMPLE 7 The motor selected for use in Example 5 is a single phase 230 V induction motor with an output of 1 kW and a power factor of 0.85. Calculate (a) the current drawn from the supply and (b) the correct size of BS EN 60898 type C protective device.

(a)

$$P = UI \times p.f.$$

$$I_L = \frac{P}{U_L \times p.f.}$$

$$= \frac{1000}{230 \times 0.8}$$

$$= 5.43 \, A$$

(b) Protective device chosen would be a 6 A type C BS EN 60898.

EXERCISE 14

1. Calculate the full-load current of each of the motors to which the following particulars refer:

	Power output (kW)	Phase	Voltage	Efficiency (%)	Power factor
(a)	5	1	230	70	0.7
(b)	3	1	250	68	0.5
(c)	15	3	400	75	0.8
(d)	6	1	230	72	0.55
(e)	30	3	400	78	0.7
(f)	0.5	1	230	60	0.45
(g)	8	3	400	65	0.85
(h)	25	3	440	74	0.75

2. You are required to record the input to a single-phase a.c. motor in kW and in kVA. Make a connection diagram showing the instruments you would use.

   A 30 kW single-phase motor delivers full-load output at 0.75 power factor. If the input is 47.6 kVA, calculate the efficiency of the motor.

3. A single-phase motor develops 15 kW. The input to the motor is recorded by instruments with readings as follows: 230 V, 100 A and 17 590 W.
   Calculate the efficiency of the motor and its power factor. Draw a diagram of the connections of the instruments. Account for the energy lost in the motor.

4. Make a diagram showing the connections of a voltmeter, an ammeter and a wattmeter, in a single-phase a.c. circuit supplying power to a motor.

   The following values were recorded in a load test of a single-phase motor. Calculate the efficiency of the motor and its power factor:

   Voltmeter reading 230 V

   Ammeter reading 75 A

   Wattmeter reading 13 kW

   Mechanical output 10 kW

5. (a)   What is power factor?

   (b)   Why is a.c. plant rated in kVA? Illustrate your answer by comparing the load on circuit cables of (i) a 10 kW d.c. motor and (ii) a 10 kW single-phase a.c. motor at the same supply, operating at a power factor of 0.7.

6. A single-phase motor drives a pump which raises 500 kg of water per minute to the top of a building 12 m high. The combined efficiency of the pump and motor is 52%, the supply voltage is 230 V and the power factor is 0.45. Calculate the supply current.

7. The output of a motor is 75 kW; the input is 100 kW. The efficiency of the motor is
   (a) 13.3%   (b) 7.5%   (c) 1.33%   (d) 75%

8. The efficiency of a motor is 80%. The input power when its output is 24 kW is
   (a) 30 kW   (b) 19.2 kW   (c) 192 kW   (d) 300 kW

Application of diversity factors

In the majority of installations, if the rating (I_N) of the protective devices contained within the distribution board were added together, the total current would, in most cases, exceed the rating of the supply fuse often quite considerably. Surprisingly the main fuse very rarely 'blows'.

If we consider a typical installation it is quite easy to see why this scenario exists.

It is very unusual in a domestic situation for all of the installed loads to be in use continuously. To minimize the size of cables and other equipment, we may apply diversity factors to certain circuits to obtain an assumed maximum demand for an installation.

Appendix 1 of the *On-Site Guide* gives information on the calculation of assumed current demand and allowances for final circuit diversity.

It should be remembered that diversity is not a precise calculation and in many instances experience and knowledge could be used instead of the tables which are a good guide only.

EXAMPLE I   A domestic premises with a 230 V 50 Hz supply protected by a 100 amp BS 1361 supply fuse has the following loads.

> Shower 10 kW protected by 45 A device
>
> Cooker 12 kW (13A socket outlet in control unit)
>   45 A device
>
> 2 × 32 amp ring final circuits
>
> 1 × 20 amp radial circuit serving socket outlets

1×16 amp immersion heater circuit

2 lighting circuits, each with 10 outlets
  (min 100 W per lamp Table 1A *On-Site Guide*)
  2 kW protected by 2 No. 6 A devices.

If we total up the current rating of the protective devices it can be seen that we have a potential current of:

$$45 + 45 + 32 + 32 + 20 + 16 + 6 + 6 = 202\,\text{A}$$

If we use Table 1B from the *On-Site Guide* to apply diversity factors we can find the assumed current.

Shower: no diversity Row 5 of table.

$$\frac{10\,000}{230} = 43.47\,\text{A}$$

Cooker: row 3 of table.

$$\frac{12\,000}{230} = 52\,\text{A}$$

From Table 1B, use first 10 A and 30% of remainder
$42 \times 30\% = 12.6\,\text{A}$

Enter on calculator $42 \times 0.3 = 12.6$

Total of $10 + 12 = 22$

Add 5 A for socket outlet $= 27\,\text{A}$

Power circuits: row 9 of table

$2 \times 32\,\text{A}$ and 1×20 standard arrangement circuit.

32 A plus 40% of 52 A ($52 \times 40\% = 20.8\,\text{A}$)

Total $= 52.8\,\text{A}$

Immersion heater: row 6 of table

No diversity $= 16\,\text{A}$

Lighting: row 1 of table

$$\frac{2000}{230} = 8.69 \times 66\% = 5.73\,\text{A}$$

Enter on calculator $8.69 \times 0.66 =$

Total assumed demand is now:

$$43 + 27 + 52 + 16 + 5 = 143\,\text{A}.$$

In this case as in many other cases, the assumed maximum demand is still greater than the supply fuse. You will find that the supply fuse has been in place for many years and never been a problem and should not give rise for concern. This is because the diversity allowed is usually quite a conservative figure.

The greatest problem with this situation is that the switched disconnector is usually $100\,\text{A}$. In these cases consideration should be given to using a split load board with possibly the power circuits on an RCD protected side of the board.

EXAMPLE 2   A retail jewellers has the following connected load supplied at $230\,\text{V}$ $50\,\text{Hz}$:

Direct heaters $2 \times 2\,\text{kW}$, $3 \times 1.5\,\text{kW}$, $1 \times 1.0\,\text{kW}$

Cooker $5\,\text{kW}$ (cooker unit has socket outlet)

Water heating (thermostatic) $3\,\text{kW}$

Socket outlets $30\,\text{A}$ ring circuit

Shop and window lighting $2.5\,\text{kW}$ total.

Determine the assumed maximum demand.

Assumed maximum demand (A.M.D.) using *IEE On-Site Guide* Table 1B (small shop premises):

Heaters

$$2 \times 2\,\text{kW} + 3 \times 1.5 + 1.0 = \frac{2000}{230} + \frac{7500}{230} \times \frac{75}{100}$$

$$= 8.7\,\text{A} + 24.46\,\text{A} = 33.16\,\text{A}$$

Cooker $\dfrac{5000}{230} = 21.7\,\text{A}$

Socket at cooker control 5 A

Water heater $\dfrac{3000}{230} = 13.04\,\text{A}$ (no diversity)

Socket outlets 30 A

Lighting $\dfrac{2500}{230} = 10.9 \times \dfrac{90}{100} = 9.8\,\text{A}$

Assumed maximum demand $= 33.16 + 21.7 + 5 + 13.04$
$$+\ 30 + 9.8$$
$$= 112.7\,\text{A}$$

In this case there may be separate main control for associated circuits. Heating and shop window lighting may be on time switch/contactor controls with individual distribution boards and switch-disconnectors.

EXAMPLE 3   A small hotel supplied at 400/230 V 50 Hz has the following connected load:

100 lighting outlets

50×13 A socket outlets on 6×30 A ring circuits

30×1 kW convection heaters on coin-operated metres

16 kW thermal storage central heating boiler

Cooking equipment – 2×14 kW cookers, 1×4 kW hot cupboard, 3×2 kW fryers, 4×600 W microwave ovens plus 5 kW machines.

Determine the assumed maximum demand.

Using Tables 1A and 1B in the *IEE On-Site Guide*:

100 lighting points = $100 \times 100\,\text{W} = 10\,\text{kW}$

so A.M.D. $= \dfrac{10\,000}{230} \times \dfrac{75}{100} = 32.6\,\text{A}$

$6 \times 30\,\text{A}$ ring circuits:

$$\text{A.M.D. is } 30\,\text{A} + \frac{150 \times 50}{100} = 105\,\text{A}$$

$30 \times 1\,\text{kW}$ convection heater:

$$\text{A.M.D. is } \frac{1000}{230} + \frac{29\,000 \times 50}{230 \times 100} = 4.35\,\text{A} + 63\,\text{A} = 67.35\,\text{A}$$

$16\,\text{kW}$ thermal storage:

$$\text{A.M.D. is } \frac{16\,000}{230} = 69.56\,\text{A} \quad \text{(no diversity)}$$

Cooking equipment

$$\frac{14\,000}{230} + \frac{14\,000 \times 80}{230 \times 100} + \frac{4000 + 6000 + 2400 + 5000 \times 60}{230 \times 100}$$

$$= 60.9\,\text{A} + 48.7\,\text{A} + 45.4\,\text{A} = 155\,\text{A}$$

Assumed maximum demand

$$= 32.6\,\text{A} + 105\,\text{A} + 67.35\,\text{A} + 69.56\,\text{A} + 155\,\text{A}$$

$$= 429.51\,\text{A}$$

Assuming that load is balanced over the three-phase supply then load would be approximately 143 A per phase.

EXERCISE 15

1. A bungalow is supplied at 230 V 50 Hz and has the following connected load:

   18 ceiling mounted lighting outlets

   $12 \times 2\,\text{A}$ socket outlets for local luminaires

3 × 30 A socket outlet ring circuits

1 × 10 kW cooker (control unit without socket outlet)

1 × 5.5 kW hob unit

10 kW of thermal storage space heating

1 × 3 kW immersion heater (thermostat controlled)

1 × 8 kW shower unit.

Determine the assumed maximum demand, and comment upon the magnitude of this.

2. A ladies hairdressing salon is supplied at 230 V 50 Hz and has the following connected load:

4 kW thermal storage space heating

6 × 3 kW under-sink instantaneous water heater

2 × 30 A socket outlet ring circuits

2 kW of shop lighting

2 × 500 W tungsten-halogen shop front luminaires.

Determine the assumed maximum demand and comment upon its magnitude.

3. A country hotel is at present supplied at 230 V 50 Hz and is to be rewired employing the following installation circuits:

Lighting: four floors each having 1000 W loading

Heating: three upper floors each having 6 × 1 kW convection heaters; ground floor 3 × 3 kW and 3 × 2 kW convection heaters

Socket outlets: 4 × 30 A ring circuits

Cooking appliances: 1 × 10 kW cooker, 1 × 6 kW hob unit, 4 kW of assorted appliances (cooker control without socket outlet)

Outside lighting 3 × 500 W tungsten halogen floodlights.

Determine the assumed maximum demand and comment upon its magnitude.

4. An insurance office is supplied at 400/230 V 50 Hz and has the following connected load:

4 × 30 A socket outlet circuits for computer use

1 × 30 A socket outlet circuit for general use

1.5 kW of fluorescent lighting

1.0 kW of tungsten lighting

1 × 6 kW cooker

2 × 600 W microwave cookers

2 × 3 kW instantaneous type hand washers

2 × 2 kW hand dryers.

Determine the assumed maximum demand.

Cable selection

Table 54G in BS 7671 gives the minimum size for circuit protective conductors. This is a simple table to use, however, it is often useful to use a smaller c.p.c. than that required by Table 54G. On larger cables, cost is a major factor as is space. If it is required to select a size of conductor smaller than is given in the table, a calculation must be carried out to ensure the conductor temperature will not rise above its final limiting temperature under fault conditions, this is called the adiabatic equation. (Final limiting temperature for conductors can be found in Table 43A.)

The phase and c.p.c. must also meet Z_s requirements for the circuit.

EXAMPLE I   The design current (I_b) for a circuit is 38 A.

The current carrying capacity of cable has been calculated and the circuit is to be wired in 70°C thermoplastic singles, 6 mm² live conductors and 1.5 mm² c.p.c. it is protected by a BS 60898 40 A type B circuit breaker. Supply is 230 V TNS with a Z_e of 0.38 Ω, circuit is 28 m long.

Calculate Z_s and requirements and thermal constraint.

The resistance of the phase and c.p.c. for this circuit must now be calculated and then compared with the values given in Table 41B2 in BS 7671.

From Table 9A *On-Site Guide*, it will be seen that the $r_1 + r_2$ for 6 mm²/1.5² copper is 15.2 mΩ per metre at a temperature of 20°C.

The cable resistance given in Table 9A is at 20°C and the Z_s for protective devices given in the tables in BS 7671 is for cables at their operating temperature of 70°C.

The cable resistance must be adjusted by calculation to allow for the increase in resistance due to the rise in temperature. The resistance of the copper conductor will increase 2% for each 5°C rise in temperature. If a cable temperature alters from 20°C to 70°C, the resistance will rise by 20%.

Table 9C from the *On-Site Guide* gives multipliers to correct the resistance of conductors at the maximum operating temperature depending on the type of cable insulation and how the c.p.c. is installed. For 70°C, thermoplastic (p.v.c.) multicore cable it can be seen that a multiplier of 1.20 must be used (if you multiply by 1.20 a value will increase by 20%).

$$\frac{m\Omega \times L \times 1.2}{1000} = R_1 + R_2$$

$$\frac{15.2 \times 28 \times 1.2}{1000} = 0.51\,\Omega$$

$$Z_s = Z_e + (R_1 + R_2)$$

$$Z_s = 0.38 + 0.511 = 0.89 \quad \text{(two decimal places is acceptable)}$$

$$Z_s = 0.89\,\Omega$$

Compare this with the maximum value given for Z_s in Table 41B2. The calculated value must be less than the tabulated value.

Max Z_s for 40 A type B mcb is 1.2 Ω. Therefore our circuit Z_s at 0.89 Ω is acceptable.

Now the prospective earth fault current must be calculated, it is important that the open circuit voltage of supply U_{Oc} is used. The calculation is

$$\frac{U_{Oc}}{Z_s} = I_f$$

$$\frac{240}{0.89} = 269.4\,\text{A}$$

Prospective earth fault current is 269.4 A (0.269 kA).
The calculation for thermal constraint should now be carried out.

From Figure 3.4 in appendix 3 of BS 7671, the disconnection time for a 40 A BS 60898 type B device can be found. From the chart on the top right of the page, it can be seen that for a 40 A device to operate within 0.1 seconds a minimum current of 200 A is required.

Regulation 543-01-03 gives the formula that must be used for thermal constraint.

$$\text{Adiabatic equation is } S = \frac{\sqrt{I^2 \times t}}{k}$$

S is the minimum permissible size of c.p.c.

I is the earth fault current

t is the time in seconds

k is the value given from Table 54C for 70°C cable

$$\frac{\sqrt{269 \times 269 \times 0.1}}{115} = 0.73 \, \text{mm}^2$$

Enter into calculator $269 \, X^2 \times 0.1 = \sqrt{} \quad = \div 115 = \text{(answer)}(0.73)$
The minimum size c.p.c. that may be used is $0.73 \, \text{mm}^2$.
Therefore $1.5 \, \text{mm}^2$ cable is acceptable.

EXAMPLE 2   A single phase 230 V circuit is to be wired in $10 \, \text{mm}^2$ phase with $1.5 \, \text{mm}^2$ c.p.c. thermoplastic 70°C copper singles cable. The protective device is a 63 amp BS 88 general purpose fuse. Z_e for the circuit is 0.23 Ω. The circuit is 36 m long and has a maximum of 5 seconds disconnection time.

Calculate to find actual Z_s and for thermal constraints.

From Table 9A *On-Site Guide* $10 \, \text{mm}^2$ (r_1) has a resistance of $1.83 \, \text{m}\Omega/\text{m}$ and $1.5 \, \text{mm}^2$ (r_2) has a resistance of $12.10 \, \text{m}\Omega/\text{m}$. Resistance of cable is $1.83 + 12.10 = 13.93 \, \text{m}\Omega$ per metre. Therefore, 36 metres will have resistance of:

$$\frac{13.93 \times 36 \times 1.2}{1000} = 0.6 \, \Omega \quad \text{(remember multiplier 1.2 for temperature correction)}$$

Calculate actual Z_s.

$$Z_s = Z_e + (R_1 + R_2)$$

$$0.23 + 0.6 = 0.83\,\Omega$$

$$Z_s = 0.83\,\Omega$$

Compare this value with maximum permissible Z_s from Table 41D in BS 7671.

Maximum permissible Z_s for a 63 A BS 88 is $0.86\,\Omega$. This will be fine as the calculated Z_s is $0.83\,\Omega$.

Now we must calculate maximum earth fault current:

$$\frac{U_{Oc}}{Z_s} = I_f$$

$$\frac{240}{0.83} = 289.91\,A$$

Now use I_f to calculate disconnection time using Figure 3.3B in BS 7671 as follows.

Along the bottom line move to the right until a vertical line matching a current of 290 A is found, follow the line up vertically until it crosses the thick black line for the 63 A fuse. In line with this junction move across to the left-hand side to find the disconnection time, which will be 4 seconds.

Now look at Table 54C of BS 7361 to find the value K for the protective conductor.

A 70°C thermoplastic cable with a copper conductor has a K value of 115.

Now carry out the adiabatic equation to ensure that c.p.c. is large enough.

On calculator enter $289.91 X^2 \times 4 = \sqrt{} = \div\, 115 = (5.04)$

$$\frac{\sqrt{I^2 \times t}}{115} = s \qquad \frac{\sqrt{289.91^2 \times 4}}{115} = 5.04\,\text{mm}^2$$

This shows that the c.p.c. is too small.

The same calculation should now be carried out using $2.5\,\text{mm}^2$ c.p.c.

From Table 9A in the *On-Site Guide*, $2.5\,\text{mm}^2$ has a resistance of $7.41\,\text{m}\Omega/\text{m}$.

$10\,\text{mm}^2/2.5\,\text{mm}^2$ has a resistance of $1.83 + 7.41 = 9.24\,\text{m}\Omega/\text{m}$

$$\frac{9.24 \times 36 \times 1.2}{1000} = 0.399\,\Omega$$

$$I_f = \frac{240}{0.399} = 601.5\,\text{A}$$

Now check disconnection time in Figure 3.4, Table 3 BS 7671.
Disconnection time is now 0.2 seconds.
Use adiabatic equation

$$\frac{\sqrt{601.5^2 \times 0.2}}{115} = 2.33\,\text{mm}^2$$

This proves that $2.5\,\text{mm}^2$ c.p.c. can be used.

EXERCISE 16

1. Calculate the $R_1 + R_2$ of $23\,\text{m}$ of the copper conductors in a $2.5\,\text{mm}^2/1.5\,\text{mm}^2$ thermoplastic twin and c.p.c. cable at $20°\text{C}$.

2. Calculate the resistance of the conductors in question 1, at their operating temperature of $70°\text{C}$.

3. A circuit is to be wired in $70°\text{C}$ thermoplastic $6\,\text{mm}^2/2.5\,\text{mm}^2$ copper cable and is $18\,\text{m}$ long, the Z_e for the circuit is $0.8\,\Omega$. Calculate the Z_s for the circuit at its maximum operating temperature.

4. The circuit above is protected by a BS 3036 semi-enclosed fuse with a disconnection time of 5 seconds. Will the circuit comply with the requirements of BS 7671.

5. A circuit is wired in thermoplastic copper $4\,\text{mm}^2$ phase with a $1.5\,\text{mm}^2$ c.p.c. it has a calculated Z_s of $1.14\,\Omega$ at $70°\text{C}$. The circuit is protected by a $30\,\text{A}$ BS 1361 fuse with a maximum disconnection time of 0.4 seconds. Will this cable comply with the requirements for the required (a) disconnection time and (b) thermal constraints?

6. If a circuit was wired in 90°C thermosetting cable with copper conductors, and had a calculated fault current of 645 A with a disconnection time of 1.5 seconds, calculate using the adiabatic equation the smallest permissible size c.p.c.

7. A circuit is required to supply a 60 A load, it is to be installed in trunking using 70°C thermoplastic (p.v.c.) singles cables with copper conductors. The circuit will be protected by a BS 88 fuse, the trunking will contain two other circuits and will be fixed using saddles to a brick wall in an ambient temperature of 35°C. Maximum permissible voltage drop for this circuit is 6 V. The circuit is 27 m long, supply is 230 V TNC-S with a Z_e of 0.35 Ω. Disconnection time is 5 seconds maximum. Calculate (a) the minimum size phase and (b) c.p.c. conductors required.

VOLTAGE DROP AND CABLE SELECTION CALCULATIONS

These calculations are fully explained in Volume 1, these are additional questions for revision.

Voltage drop calculations

The voltage drop in cable conductor(s) is directly proportional to the circuit current and the length of cable run.

Voltage drop

$$= \frac{\text{current (A)} \times \text{length of run (m)} \times \text{millivolt drop per A/m}}{1000}$$

(Note division by 1000 to convert millivolts to volts.)

Note BS 7671 Requirement 525-01-02 limits the voltage drop permitted between the origin of the installation and the terminals of a load to 4% of the nominal supply voltage. For a single-phase 230 V supply this equates to 9.2 V, and for 400 V three-phase supply to 16 V.

EXAMPLE 1   A 3 kW 230 V 50 Hz single-phase motor has an efficiency of 70% and works at a power factor of 0.6. It is connected to its starter by single-core p.v.c.-insulated cables (copper conductors) drawn into steel conduit (method 3); the length of run is 25 m. The voltage drop in the cables must not exceed 6 V. Assume an ambient temperature of 35°C and protection by BS 88 fuses.

Circuit details:

Motor circuit, starter will offer overload protection.

Ambient temperature 35°C *so* C_a is 0.94.

Using BS 88 (Gm) fuses so C_r is 1.

Output = 3 kW

$$\text{Input} = 3000 \times \frac{100}{70}$$

$$= 4285.7 \, \text{W}$$

$$P = U \times I \times p.f.$$

$$4285.7 = 230 \times I \times 0.6$$

$$I_b = \frac{4285.7}{230 \times 0.6}$$

$$= 31.1 \, \text{A}$$

Minimum BS 88 fuse rating (I_n) say 40 A (allows for moderate overcurrent at starting). Starter will offer overload protection (see BS 7671 requirements 435-01-01 and 552-01-02).

Correction factors applying:

$$C_a \text{ is } 0.94 \, (35°C)$$

$$C_r \text{ is } 1 \, (\text{BS 88 fuses})$$

Thus minimum current rating:

$$(I_t) = \frac{40}{0.94 \times 1}$$

$$= 42.55 \, \text{A}$$

Using BS 7671 Table 4D1A or *IEE On-Site Guide* Table 6D1, from column 4 select $10\,\text{mm}^2$ cables ($57\,\text{A}$) and using BS 7671 Table 4D1B or *IEE On-Site Guide* Table 6D2, column 3, read mV/A/m value for $10\,\text{mm}^2$ cables as $4.4\,\text{mV/A/m}$

$$\text{Volts drop in 25 m} = \frac{31.1 \times 25 \times 4.4}{1000}$$

$$= 3.42\,\text{V}$$

Thus $10\,\text{mm}^2$ cables will be suitable.

EXAMPLE 2

(a) An industrial process heater of rating $16\,\text{kW}$ is fed at $400\,\text{V}$ $50\,\text{Hz}$. Three-phase four-wire is to be installed in a factory using a p.v.c.-insulated, non-armoured, copper conductors multicore cable. Length of run is $25\,\text{m}$ clipped direct to a wall; assume a maximum ambient temperature of $35°\text{C}$ and protection by BS 3036 fuses.
(b) If the BS 3036 fuses were replaced by BS 88 (Gg) fuses what would be the effect on cable current rating?

Circuit details:

As it is a heater p.f. is unity.

Ambient temperature $35°\text{C}$ so C_a is 0.94.

Using BS 3036 fuses so C_r is 0.725.

$$\text{Current demand } I_b = \frac{16\,000}{\sqrt{3} \times 400}$$

$$= 23.1\,\text{A}$$

(a) Select as I_n $30\,\text{A}$ BS 3036 fuses. Thus minimum current rating is

$$I_t = \frac{30}{0.94 \times 0.725}$$

$$= 44\,\text{A}$$

Using BS 7671 Table 4D2A or *IEE On-Site Guide* Table 6E1, from column 7 select $10\,\text{mm}^2$ cables (57 A) and using BS 7671 Table 4D2B or *IEE On-Site Guide* Table 6E2, column 4, read mV/A/m value for $10\,\text{mm}^2$ cables as $3.8\,\text{mV/A/m}$.

$$\text{Volts drop in 25 m} = \frac{23.1 \times 25 \times 3.8}{1000}$$

$$= 2.19\,\text{V}$$

Thus $10\,\text{mm}^2$ cables will be suitable.

(b) Select as I_n 25 A BS 88 fuses. In this case C_r is 1. Thus minimum current rating is

$$I_\text{t} = \frac{25}{0.94 \times 1}$$

$$= 26.6\,\text{A}$$

Using BS 7671 Table 4D2A or *IEE On-Site Guide* Table 6E1, from column 7 select $4\,\text{mm}^2$ cables (32 A) and using BS 7671 Table 4D2B or *IEEE On-Site Guide* Table 6E2, column 4, read mV/A/m value for $4\,\text{mm}^2$ cables as $9.5\,\text{mV/A/m}$.

$$\text{Volts drop in 25 m} = \frac{23.1 \times 25 \times 9.5}{1000}$$

$$= 5.49\,\text{V}$$

Thus $4\,\text{mm}^2$ cables will be suitable.

EXAMPLE 3   A p.v.c. trunking containing single-core p.v.c.-insulated distribution cables (copper conductors) is to be run 30 m from the main switchgear of an office building to supply a new $400/230\,\text{V}$ T.P. & N distribution fuseboard. The balanced load consists of 18 kW of discharge lighting. The main and local distribution boards employ fuses to BS 88 (Gg) Part 2. The voltage drop in the distribution cables must not exceed 6 V and the ambient temperature is anticipated to be $30°\text{C}$.

For the distribution cables, establish and state the

(i)   design current I_b

(ii) minimum rating of fuse in the main switch fuse I_n

(iii) maximum mV/A/m value

(iv) minimum current rating I_t

(v)   minimum cross-sectional area of the live conductors

(vi) actual voltage drop in the cable   (C&G)

> *Circuit details*:
>
> Discharge lighting circuit requires a multiplier of 1.8
>
> (*IEE On-Site Guide*, Appendix 1).
>
> Ambient temperature 30°C so C_a is 1.
>
> Using BS 88 fuses so C_r is 1.
>
> Cable voltage drop limitation of 6 V.
>
> Cables in trunking to method 3.

(i)   Design current $I_b = \dfrac{18 \times 10^3 \times 1.8}{\sqrt{3} \times 400}$

$$= 46.77 \, \text{A}$$

(ii) Minimum BS 88 fuse rating is 50 A.

(iii) Maximum mV/A/m value $= \dfrac{6 \times 1000}{46.77 \times 30}$

$$= 4.28 \, \text{mV/A/m}$$

(iv) Minimum current rating $= \dfrac{50}{1}$

$$= 50 \, \text{A}$$

(v)   From BS 7671 Tables 4D1A and 4D1B or *IEE On-Site Guide*

Tables 6D1 (column 5) and Table 6D2 (column 6), minimum c.s.a. of cable is $16 \, \text{mm}^2$ (68 A/2.4 mV/A/m).

(vi) Actual voltage drop in 30 m cable $= \dfrac{46.77 \times 30 \times 2.4}{1000}$

$$= 3.37 \, \text{V}$$

EXAMPLE 4 A 400 V 50 Hz three-phase extract fan has a rating of 15 kW at 0.8 p.f. lagging and is supplied from a BS 88 (Gg) Part 2 type distribution board 40 m distant. The cables are to be single-core, p.v.c.-insulated, run in steel trunking with three similar circuits. Assume an ambient temperature of 35°C and that the voltage drop in the cables is limited to 2.5% of the line voltage.

Establish the:

(i)   full load current of the motor I_L
(ii) rating of the fuses I_n
(iii) minimum current rating of cables
(iv) minimum cable c.s.a.
(v)   actual voltage drop in cables

> *Circuit details*:
>
> Extract fan circuit: low starting current.
>
> Four sets of circuit cables: C_g is 0.65.
>
> Ambient temperature: 35°C so C_a is 0.94.
>
> Using BS 88 fuses, so C_r is 1.
>
> Cable voltage drop limitation of 2.5% of 400, i.e. 10 V.
>
> Cables in trunking to method 3.

(i) As $P = \sqrt{3} U_L I_L \cos\phi$

$$15\,000 = \sqrt{3} \times 400 \times I_L \times 0.8$$
$$I_L = \frac{15\,000}{\sqrt{3} \times 400 \times 0.8}$$
$$= 27 \, \text{A}$$

(ii) Select 32 A BS 88 fuses (allowing for low starting current).

(iii) Minimum current rating of cables $= \dfrac{32}{0.94 \times 0.65}$

$$= 52.4 \, \text{A}$$

(iv) From BS 7671 Tables 4D1A and 4D1B or *IEE On-Site Guide* Tables 6D1 (column 5) and Table 6D2 (column 6) select 16 mm² (68 A/2.4 mV/A/m).

(v) Voltage drop in 40 m $= \dfrac{27 \times 40 \times 2.4}{1000}$

$= 2.6\,\text{V}$

As volts drop limitation is 4% of 400 V, i.e. 16 V, 16 mm² cable is satisfactory.

The final example illustrates the effect on the required tabulated cable rating of combined correction factors.

EXAMPLE 5   A twin and earth p.v.c.-insulated (copper conductors) cable runs between a 230 V distribution board at the origin of an installation and a 10 kW heater. The cable passes through the following environmental conditions:

(a) on its own in a switchroom with an ambient temperature of 35°C;

(b) on its own in an outdoor area with an ambient temperature of 25°C;

(c) bunched with three other cables on a wall surface in an area with an ambient temperature of 40°C;

(d) finally on its own passing through a thermally insulated wall section for a distance of 2 m, in an ambient temperature of 30°C.

Protection is by BS 3036 fuses, length of run is 60 m and the voltage drop is limited to 5.5 V.

Calculate the minimum cable rating and select suitable cable for voltage drop limitation.

Circuit details:

Heater circuit so no special restrictions.

Protection by BS 3036 fuses so C_r is 0.725.

Voltage drop limitation is 5.5 V.

Area (a)   35°C C_a is 0.94.

Area (b)   25°C C_a is 1.03.

Area (c)   40°C C_a is 0.87, C_g is 0.65.

Area (d)   30°C C_a is 1, C_i is 0.5.

Now overall correction factors are as follows:

Area (a) $0.94 \times 0.725 = 0.68$

Area (b) $1.03 \times 0.725 = 0.747$

Area (c) $0.87 \times 0.65 \times 0.725 = 0.41$

Area (d) $0.5 \times 0.725 = 0.36$ (worst area)

$$\text{Design current } I_b = \frac{10\,000}{230}$$

$$= 43.5\,\text{A}$$

Nearest BS 3036 fuse element is 45 A (BS 7671 Table 53A). Select worst area (d): C_a is 0.5.

$$\text{Minimum cable rating} = \frac{45}{0.36}$$

$$= 43.5\,\text{A}$$

From BS 7671 Table 4D2A or *IEE On-Site Guide* Table 6E1 and from BS 7671 Table 4D2B or *IEE On-Site Guide* Table 6E2 select $70\,\text{mm}^2$ (139 A) and $0.63\,\text{mV/A/m}$.

$$\text{Volts drop in 60 m} = \frac{43.5 \times 60 \times 0.63}{1000}$$

$$= 1.64\,\text{V}$$

So $70\,\text{mm}^2$ cable is satisfactory.

Obviously one should avoid running cables in hostile environments wherever possible, in this case avoiding thermal insulation and not using BS 3036 protection. Assuming that the cable grouping was unavoidable we could now use area (c) as the worst environment and in this case:

Revised circuit details:

Heater circuit so no special restrictions.

Protection by BS 88 fuses so C_r is 1.

Voltage drop limitation is 5.5 V.

Area (c) $0.87 \times 0.65 = 0.565$

Voltage drop limitation 5.5 V.

$$\text{Minimum cable rating} = \frac{45}{0.565}$$

$$= 79.6\,\text{A}$$

From BS 7671 Table 4D2A or *IEE On-Site Guide* Table 6E1 and from BS 7671 Table 4D2B or *IEE On-Site Guide* Table 6E2 select $25\,\text{mm}^2$ (90 A) and $1.75\,\text{mV/A/m}$.

$$\text{Volts drop in } 60\,\text{m} = \frac{43.5 \times 60 \times 1.75}{1000}$$

$$= 4.75\,\text{V}$$

So $25\,\text{mm}^2$ cable is satisfactory and cheaper to install than $70\,\text{mm}^2$ cable.

EXERCISE 17

1. Establish the current-carrying capacity (I_z) of a cable with a tabulated current rating (I_t) of 17.5 A when it is grouped in conduit with two other circuits in an ambient temperature of 35°C; protection is by BS 3036 fuses.

2. Calculate the actual voltage drop and the power wasted in a $25\,\text{mm}^2$ cable, 10 m long, when it carries 70 A. The listed mV/A/m for the cable is 1.8 mV.

3. The design current of a single-phase circuit is 35 A. The single-core p.v.c.-insulated cables run alone in p.v.c. conduit for a distance of 50 m through an area having an ambient temperature of 35°C (100 mm of the conduit passes through thermal insulation). The voltage drop in the circuit must not exceed 5 V. Protection is by a BS 1361 fuse. Determine the:

   (a) fuse rating

   (b) minimum cable current rating

   (c) minimum cable c.s.a.

   (d) voltage drop in the cables.

4. A supply is required to a 3 kW heater which is 25 m from a local BS 1361 distribution board. The building is fed at 230 V 50 Hz single-phase. It is proposed to employ a 2.5 mm^2 two-core and earth p.v.c.-insulated (copper conductors) cable for this circuit installed as method 1. Allowing for a 2 V drop in the cables feeding the distribution board, determine the:

   (a) design current

   (b) maximum volts drop permitted

   (c) volts drop in the cable

   (d) actual voltage at the heater.

5. A 10 kW motor having an efficiency of 60% is fed from a 220 V d.c. supply through cables 20 m long and having a listed voltage drop figure of 1.3 mV/A/m. Determine the:

   (a) design current

   (b) volts drop in the cables when the motor is fully loaded.

6. After the application of correction factors, a pair of single-core p.v.c.-insulated cables in conduit are required to carry 25 A from a distribution board to a load 90 m away. The voltage drop in the cables should not exceed 5 V. Using BS 7671 documents:

   (a) calculate the maximum mV/A/m value

   (b) select a suitable cable c.s.a.

   (c) calculate the voltage drop in the cables.

7. A 12 kW load is to be supplied from a 230 V main switch-fuse 65 m distant. The voltage drop is to be limited to 2.5% of the supply voltage. Overload protection is to be provided by a BS 3036 semi-enclosed fuse. The single-core p.v.c.-insulated cables run in conduit with one other single-phase circuit. Assuming an ambient temperature of 25°C, determine with the aid of BS 7671 documents:

   (a) the design current

   (b) the fuse rating

   (c) the minimum cable current rating

   (d) the maximum mV/A/m value

(e) the selection of a suitable cable c.s.a.

(f) the voltage drop in the circuit.

8. A single-phase load of 10 kW is to be supplied from a 230 V distribution board 120 m distant. Overload protection is to be by BS 88 (Gg) Part 2 fuses. The twin with earth, p.v.c.-insulated cable is clipped with three similar cables as BS 7671 method 1 in an ambient temperature of 25°C. Voltage drop in the cables should not exceed 5 V. Determine with the aid of BS 7671 documents the:

(a) design current

(b) fuse rating

(c) minimum cable current rating

(d) maximum mV/A/m value

(e) minimum cable c.s.a.

(f) voltage drop in the cables.

9. A 400/230 V 50 Hz T.P. & N distribution board is to be installed in a factory to feed 11 kW of mercury vapour lighting. Due to the adverse environmental conditions, it is intended to use p.v.c. conduit to contain the single-core p.v.c.-insulated cables (copper conductors). The total length of the run from the main switchboard is 50 m. To provide earthing protection it is intended to draw a 4 mm^2 single-core p.v.c.-insulated cable (copper conductors) into the conduit. The following details apply to the installation:

(i) an ambient temperature of 35°C

(ii) BS 88 (Gg) Part 2 fuse protection throughout

(iii) voltage drop in the cables must not exceed 8.5 V

(iv) the BS 88 fuses satisfy the requirements of BS 7671 Requirement 434-03-03.

Establish the:

(a) design current

(b) rating of fuses in the main switchboard

(c) minimum current rating of live conductors

(d) maximum mV/A/m value of live conductors

(e) minimum cross-sectional area of live conductors

(f) actual voltage drop in submain cables

(g) size of p.v.c. conduit to be used, assuming that one section of the run involves one right-angle bend in 8 m.   (C&G)

10. A single steel trunking is to be run from a 400/230 V 50 Hz main switchboard to feed three SP & N lighting distribution boards containing Type 2 BS 3871 miniature circuit breakers, sited at 5, 12 and 20 m distances. Each distribution board feeds 5 kW of mercury vapour lighting. The following details apply to the installation:

   (i) ambient temperature in the area is 25°C

  (ii) protection at the main switchboard is by BS 88 fuses

 (iii) single-core p.v.c.-insulated (copper conductors) cables are to be employed

 (iv) voltage drop in the distribution cables must not be greater than 3.5 V.

Establish the:

(a) design current

(b) maximum mV/A/m value permitted

(c) fuse rating at the main switchboard

(d) minimum cable current rating

(e) minimum cross-sectional area of the distribution cables

(f) voltage at each distribution board.   (C&G)

11. A 12 m length of two-core and earth, p.v.c.-insulated cable is clipped to a surface as BS 7671 method 1. The cable feeds a load of 4 kW at 230 V 50 Hz a.c. The following details apply to the installation:

   (i) power factor of the load is 0.8 lagging

  (ii) ambient temperature of 20°C

 (iii) protection by a BS 88 (Gg) Part 2 fuse

 (iv) cable voltage drop not to exceed 5 V.

Determine the:

(a) design current

(b) rating of the fuse

(c) minimum cable current rating

(d) maximum mV/A/m value

(e) minimum cable cross-sectional area

(f) actual voltage drop in the cable at full load.

12. A 4.5 kW single-phase load in a factory is to be supplied from the 400/230 V 50 Hz suppliers' main switchboard 40 m distant, using two-core and earth, p.v.c.-insulated cable. The power factor of the load is 0.7 lagging and the cable route is through an ambient temperature of 30°C. Protection is by BS 88 (Gg) Part 2 fuses.
Determine the:

(a) design current

(b) permissible voltage drop in circuit

(c) minimum fuse rating

(d) minimum cable current rating

(e) maximum mV/A/m value

(f) minimum cable cross-sectional area

(g) actual voltage drop in the cable at full load.

13. A 230 V 50 Hz 8 kW electric shower unit is to be installed in an industrial premises using a two-core and earth, p.v.c.-insulated cable, 20 m in length. The ambient temperature is 30°C. Protection is by a BS 1361 fuse in a distribution board, the cable volts drop should not exceed 2 V.
Determine the:

(a) design current I_b

(b) fuse rating I_n

(c) required cable current rating I_t

(d) required cable c.s.a.

(e) actual volts drop in the cable.

14. A 25 kW 400 V 50 Hz three-phase motor operates at 0.85 p.f. lagging on full load. The p.v.c.-insulated single-core cables run together for a distance of 10 m with two similar circuits through a trunking to a circuit breaker

distribution board. Assume that the circuit breaker is selected to have an operating value of not less than 1.5 times the motor full-load current, the ambient temperature is 35°C and the voltage drop in the cables should not exceed 10 V.

Determine the:

(a) design current

(b) setting of circuit breaker

(c) minimum cable current rating

(d) maximum mV/A/m value

(e) minimum cable c.s.a.

(f) actual volts drop in the cable.

15. The voltage-drop figure for a certain cable is $2.8\,\mathrm{mV/A/m}$. The actual drop is 50 m run of this cable when carrying 45 A is:

(a) 1.2 V   (b) 6.3 V   (c) 0.1 V   (d) 10 V

16. The voltage drop allowed in a certain circuit is 6 V. The length of run is 35 m. The cable used has a voltage-drop figure of $7.3\,\mathrm{mV/A/m}$. Ignoring any correction factors, the maximum current which the cable can carry is:

(a) 15 A   (b) 23.5 A   (c) 41 A   (d) 43.8 A

17. A circuit is given overload protection by a 30 A BS 3036 fuse. The grouping factor C_g is 0.65 and the ambient temperature factor is 0.87. The minimum current-carrying capacity of the cable should be:

(a) 73.2 A   (b) 53 A   (c) 30 A   (d) 41.3 A

18. A 10 kW 230 V a.c. motor operates at 0.75 lagging. The starter offering overload protection is set at 1.5 times the F.L.C. of the motor. Ignoring any correction factors, the minimum current-carrying capacity of the cable to the motor required is:

(a) 43.5 A   (b) 58 A   (c) 87 A   (d) 108.7 A

19. A certain cable having a tabulated current rating (I_t) of 18 A has correction factors of 1.04, 0.79 and 0.725 applied to compensate for its operating conditions. The operational current rating (I_z) for the cable is:

(a) 30.22 A   (b) 24.83 A   (c) 13.572 A   (d) 10.72 A

To prevent danger to persons, livestock and property every installation must be protected against the persistence of earth leakage currents. This is generally achieved by providing a low-impedance earth-current leakage path from the installation to the source of supply, i.e. the local distribution transformer.

The leakage path must have a low enough impedance (Z_s) to allow adequate earth leakage current to flow to 'blow' the circuit fuse or operate the circuit breaker and isolate the faulty circuit within a specified time, usually either 5 seconds or 0.4 seconds. BS 7671 gives guidance to the permissible earth-loop impedance values to meet the disconnection times and that document and the *IEE On-Site Guide* contain tables which list types of protective device and specify the maximum measured earth fault loop impedance in ohms for each rating of the specific device. Where precise disconnection times are demanded then BS 7671 Appendix 3 contains characteristic curves for fuses and circuit breakers.

Part of the earth leakage path is outside the control of an electricity consumer and its impedance (Z_e) contributes to the total value of earth loop impedance. The value of this external impedance is generally declared by the supplier and is used in the calculation of the 'prospective' Z_s. The declared value of Z_e, however, can never be a precise value because of the supplier's service conditions at the moment of earth fault; thus the actual value of Z_s must always be measured by earth loop impedance test instruments at various points within an installation when the particular circuit is energized and is under test-load conditions.

For the estimation of prospective earth-loop impedance values we may however regard Z_e as an empirical or estimated value when assessing the permitted value of the installation's internal cable impedance (or resistance) value.

The internal cable 'impedance' will be determined by the cross-sectional area and resistance (R_1) of the circuit's phase conductor and that of the circuit's protective conductor (R_2) from the origin of the installation to the point of connection to current-using equipment when the circuit is energized and the cables are working in their maximum operating temperature.

To predict the actual disconnection time for a earth leakage fault condition we may employ characteristic curves of the protective devices, i.e. fuses and circuit breakers. Appendix 3 of BS 7671 gives specimens of such curves.

Note For all the following examples and exercises p.v.c.-insulated copper conductors are to be employed.

EXAMPLE I An installation is being carried out and it is necessary to estimate the prospective total earth loop impedance of circuits. In order to arrive at a typical value, a lighting circuit is chosen as that is likely to have a fairly high impedance value. The circuit is to be wired in $1.5\,\text{mm}^2$ twin and earth cable (assume a $1.0\,\text{mm}^2$ protective conductor); the length of cable is $18\,\text{m}$. The declared value of Z_e is $0.35\,\Omega$. Circuit protection at the origin of the installation (consumer unit) is by a BS 1361 5 A fuse.
(a) Establish conformity with BS 7671 requirements.
(b) Establish from BS 7671 Appendix 3, the actual disconnection time.

This is a fixed-equipment circuit; five-second disconnection time.

From Tables 9A and 9C (*IEE On-Site Guide*) $R_1 + R_2$ of $1.5\,\text{mm}^2/1.0\,\text{mm}^2$ conductors is $30.2\,\text{m}\Omega/\text{m} \times 1.20$.

Thus $R_1 + R_2$ of $1.5\,\text{mm}^2/1.0\,\text{mm}^2$ conductors $18\,\text{m}$ long will be

$$\frac{18 \times 30.2 \times 1.20}{1000} = 0.65\,\Omega$$

and

$$Z_s = 0.35 + 0.65$$

$$= 1.0\,\Omega$$

(a) From Tables 2C and 2E (*IEE On-Site Guide*) maximum measured earth fault loop impedance is $13.68 \times 1.06\,\Omega$, i.e. $14.5\,\Omega$, thus the estimated value of the earth fault loop impedance for this circuit is acceptable.

(b) Actual disconnection time

$$\text{Prospective earth fault current} = \frac{230}{1}$$

$$= 230\,\text{A}$$

From Appendix 3, Table 3.1, the circuit disconnects in less than 0.1 second; we may say that the fuse operates instantaneously.

EXAMPLE 2 A commercial cooker circuit is fed by $16\,\text{mm}^2$ single-core p.v.c.-insulated cable with a $6\,\text{mm}^2$ single-core p.v.c.-insulated protective conductor cable from a BS 88 (Gg) Part 2 type fuseboard (40 A fuse) at the origin of the installation; length of cables within p.v.c. conduit is 35 m. Assume a tested Z_e value of $0.7\,\Omega$.

(a) Establish conformity with BS 7671 requirements regarding the value of Z_s.

(b) Establish from BS 7671 Appendix 3 the actual disconnection time.

(a) This is a fixed-equipment circuit; five-second disconnection time. From Tables 9A and 9B (*IEE On-Site Guide*) $R_1 + R_2$ of $16\,\text{mm}^2/6\,\text{mm}^2$ conductors is $4.23\,\text{m}\Omega/\text{m} \times 1.20$.

Thus $R_1 + R_2$ of $16\,\text{mm}^2/6\,\text{mm}^2$ conductors 15 m long will be

$$\frac{35 \times 4.23 \times 1.20}{1000} = 0.178\,\Omega$$

and

$$Z_s = 0.7 + 0.178$$

$$= 0.878\,\Omega$$

From Tables 2B and 2E (*IEE On-Site Guide*) maximum measured earth fault loop impedance is $1.13 \times 1.06\,\Omega$, i.e. $1.2\,\Omega$, thus the estimated value of the earth fault loop impedance for this circuit is acceptable.

(b) Actual disconnection time

$$\text{Prospective earth fault current} = \frac{230}{0.878}$$

$$= 261 \, \text{A}$$

From Appendix 3, Table 3.3B, the circuit disconnects in 0.6 seconds.

EXAMPLE 3　A cable feeds a single-phase electric pump and lighting point in an external building of a factory, the total length of the twin with earth $4 \, \text{mm}^2 / 2.5 \, \text{mm}^2$ p.v.c.-insulated cable is 30 m, protection is by a BS 3871, type 3, 30 A mcb in a distribution board at the suppliers' intake position. The tested value of Z_s at the intake position is 0.35 Ω.

This is a fixed-equipment circuit but is in adverse conditions, thus a 0.4 s disconnection time applies (BS 7671 requirement 471-08-03).

From Tables 9A and 9B (*IEE On-Site Guide*) $R_1 + R_2$ of $4 \, \text{mm}^2 / 2.5 \, \text{mm}^2$ conductors is 12.02 mΩ/m × 1.20.

Thus $R_1 + R_2$ of $4 \, \text{mm}^2 / 2.5 \, \text{mm}^2$ conductors 30 m long will be

$$\frac{30 \times 12.02 \times 1.20}{1000} = 0.43 \, \Omega$$

and

$$Z_s = 0.35 + 0.43$$

$$= 0.78 \, \Omega$$

From Tables 2D and 2E (*IEE On-Site Guide*) maximum measured earth fault loop impedance is 0.64 × 1.06 Ω, i.e. 0.67 Ω, thus the estimated value of the earth fault loop impedance for this circuit is not acceptable. A residual current device in the supply to the external building will be necessary.

Note Assume that copper conductor cables are used for all questions.

1. A single-phase process heater circuit is wired in p.v.c. trunking and is to employ $6\,mm^2$ single-core p.v.c.-insulated live conductors and a $2.5\,mm^2$ protective conductor. The distance from the BS 88 (Gg) Part 2 distribution fuseboard at the main switchgear is $33\,m$, rating of fuse is $40\,A$ and ambient temperature is $25°C$. The tested Z_e value at the main switchgear is $0.3\,\Omega$.

   (a) Estimate the prospective value of Z_s.

   (b) State the maximum permissible measured Z_s value.

2. A single-phase lighting circuit in a commercial premises is wired in p.v.c. conduit employing $1.5\,mm^2$ live conductors and a $1.5\,mm^2$ protective conductor. The distance from the BS 1361 distribution fuseboard at the main switchgear is $20\,m$, rating of the fuse is $15\,A$ and ambient temperature is $30°C$. The tested Z_e value at the main switchgear is $0.45\,\Omega$.

   (a) Estimate the prospective value of Z_s.

   (b) State the maximum permissible measured Z_s value.

3. A three-phase electric motor circuit is to be wired in steel trunking and is to employ $4\,mm^2$ live conductors and the client demands that an independent $2.5\,mm^2$ protective conductor is used. The distance from the BS 88 (Gm) Part 2, distribution fuseboard at the main switchgear is $10\,m$, and the rating of the fuse is $10\,A$. The testing Z_s value at the distribution board is $0.45\,\Omega$, and the ambient temperature is $25°C$.

   (a) Estimate the prospective value of Z_s at the motor starter.

   (b) State the maximum permissible measured Z_s value.

4. A 400/230V 50Hz three-phase milling machine is to be wired in p.v.c. trunking and is to employ $6\,mm^2$ live conductors and the designer specifies that an independent $4\,mm^2$ protective conductor is to be used. The distance from the BS 88 (Gm) Part 2, distribution fuseboard at the

main factory switchgear is 18 m, and the rating of the fuse is 50 A. The tested Z_s value at the distribution board is 0.4 Ω, and the ambient temperature is 20°C.

(a) Estimate the value of Z_s at the machine's starter isolator.

(b) Assuming that the actual value of Z_s is as estimated, what earth fault current will flow in the event of a direct to earth fault at the isolator?

(c) What will be the approximate disconnection time?

5. An earth fault current of 250 A occurs in a circuit protected by a BS 88 (Gg) Part 2 32 A fuse. The disconnection time will be approximately:

   (a) 0.1 s   (b) 0.2 s   (c) 0.25 s   (d) 3 s

6. An earth fault current of 130 A occurs in a circuit protected by a BS 3036 30 A fuse. The disconnection time will be approximately:

   (a) 0.8 s   (b) 0.13 s   (c) 1 s   (d) 8 s

7. An earth fault current of 300 A occurs in a circuit protected by a BS 1361 45 A fuse. The disconnection time will be approximately:

   (a) 0.18 s   (b) 1.8 s   (c) 0.3 s   (d) 0.9 s

Lighting calculations

UNITS AND QUANTITIES

Luminous intensity is the power of light from the source measured in **candela**

Illuminance is a measure of the density of luminous flux at a surface measured in **lux** (lumens per square metre)

Luminous flux is the light emitted by a source and is measured in **lumens**

Luminance is a measure of the light reflected from a surface measured in **candela per m^2**

Luminous efficacy is the ratio of the luminous flux emitted by a lamp to the power the lamp consumes this is measured in **lumens per watt**

Quantity	Quantity symbol	Unit	Unit Symbol
Luminous intensity	I	Candela	cd
Luminous flux	Φ	Lumens	lm
Illuminance	E	Lux	lx
Luminance	L	Candela square metre	cd/m^2
Luminous efficacy		Lumens per watt	lm/W

INVERSE SQUARE LAW

When using the inverse square law, the distance used in the measurement is from the light source to a point directly below it.

When a lamp is suspended above a surface, the illuminance at a point below the lamp can be calculated:

$$\text{Illuminance } E = \frac{(I)\,\text{candela}}{\text{distance}^2}\,\text{lux}$$

$$= \frac{I}{d^2}$$

EXAMPLE 1 A luminaire producing a luminous intensity of 1500 candela in all directions below the horizontal, is suspended 4 m above a surface. Calculate the illuminance produced on the surface immediately below the luminaire.

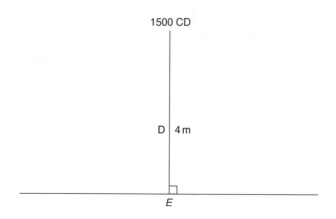

Fig. 81

$$E = \frac{I}{d^2}$$

$$= \frac{1500}{4^2}$$

$$= 93.75\,\text{lux}$$

EXAMPLE 2 If the luminaire in Example 1 is raised by 1 m, what would the new illuminance be at the point immediately below the surface?

$$E = \frac{1500}{(4+1)^2}$$

$$= \frac{1500}{5^2}$$

$$= 60 \text{ lux}$$

COSINE LAW

When using the cosine law, the distance used is from the light source measured at an angle to the point at which the lux value is required.

When a lamp is suspended above a horizontal surface, the illuminance (E) at any point below the surface can be calculated.

$$E = \frac{I}{h^2} \cos \phi$$

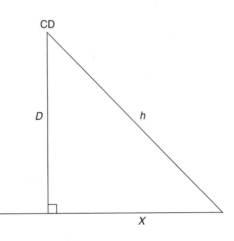

Fig. 82

To calculate h^2 (distance from lamp)

$$h^2 = d^2 + x^2$$

To calculate $\cos \phi$

$$\cos \phi = \frac{d}{h}$$

EXAMPLE I   A light source producing 1500 candela is suspended 2.2 m above a horizontal surface. Calculate the illumination produced on the surface 2.5 m away (Q)

Calculate h^2 using pythagoras.

$$h^2 = d^2 + x^2$$
$$= 2.2^2 + 2.5^2$$
$$h^2 = 11.09$$

Calculate h

$$= \sqrt{h^2}$$
$$= 3.33$$

Calculate $\cos \phi$ using pythagoras

$$= \frac{d}{h}$$
$$= \frac{2.2}{3.33}$$
$$= 0.66$$
$$E_Q = \frac{1500}{11.09} \times 0.66$$
$$= 89.26 \, \text{lux}$$

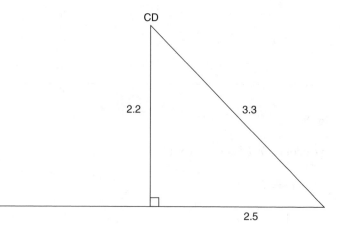

Fig. 83

EXAMPLE 2 Two lamps are suspended 10 m apart and at a height of 3.5 m above a surface (Figure 84). Each lamp emits 350 cd. Calculate.

(a) the illuminance on the surface midway between the lamps,

(b) the illuminance on the surface immediately below each of the lamps.

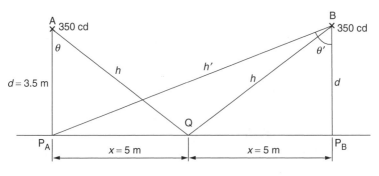

Fig. 84

(a) For one lamp, the illuminance at Q is

$$E_Q = \frac{I}{h^2} \cos \theta$$

$$= \frac{350}{3.5^2 + 5^2} \times \frac{3.5}{\sqrt{(3.5^2 + 5^2)}}$$

$$= \frac{350}{12.25 + 25} \times \frac{3.5}{\sqrt{(12.25 + 25)}}$$

$$= \frac{350}{37.25} \times \frac{350}{\sqrt{37.25}}$$

$$= 5.388 \, \text{lx}$$

The illuminance from two lamps is double that due to one lamp, since the conditions for both lamps are identical. Thus

$$\text{total illuminance at Q} = 2 \times 5.388$$

$$= 11.8 \, \text{lx}$$

(b) At P_A below lamp A, the illuminance due to lamp A is

$$E_{P_A} = \frac{I}{d^2}$$

$$= \frac{350}{3.5^2}$$

$$= 28.57 \text{ lx}$$

In calculating the illuminance at P_A due to lamp B, we have a new distance h', a new distance x', and a new angle θ' to consider.

$$x' = 2x$$

$$= 10$$

$$(h')^2 = 3.5^2 + 10^2$$

$$= 112.25$$

$$\therefore \quad (h') = 10.59$$

$$\cos \theta' = \frac{d}{h'}$$

$$= \frac{3.5}{10.59}$$

$$= 0.331$$

$\therefore$ illuminance at P_A due to lamp B is

$$E_{P_B} = \frac{350}{112.25} \times 0.331$$

$$= 1.013$$

$$\text{Total illuminance at } P_A = 28.57 + 1.013$$

$$= 29.61 \text{ lx}$$

and, as the conditions at P_B are the same as those at P_A, this will also be the illuminance below lamp B.

EXERCISE 19

1. A lamp emitting 250 cd in all directions below the horizontal is fixed 4 m above a horizontal surface. Calculate the illuminance at (a) a point P on the surface vertically beneath the lamp, (b) a point Q 3 m away from P.

2. Two luminaires illuminate a passageway. The luminaires are 12 m apart. Each emits 240 cd and is 3 m above the floor. Calculate the illuminance at a point on the floor midway between the luminaires.

3. Determine the illuminance at a point vertically beneath one of the luminaires in question 2.

4. An incandescent filament luminaire is suspended 2 m above a level work bench. The luminous intensity in all directions below the horizontal is 400 candelas.

   Calculate the illuminance at a point A on the surface of the bench immediately below the luminaire, and at other bench positions 1 m, 2 m and 3 m from A in a straight line. Show the values on a suitable diagram.   (CGLI)

5. Two incandescent filament luminaires are suspended 2 m apart and 2.5 m above a level work bench. The luminaires give a luminous intensity of 200 candelas in all directions below the horizontal. Calculate the total illuminance at bench level, immediately below each luminaire and midway between them.

6. A work bench is illuminated by a luminaire emitting 350 cd in all directions below the horizontal and mounted 2.5 m above the working surface.

   (a) Calculate the illuminance immediately below the luminaire.

   (b) It is desired to increase the illuminance by 10%. Determine two methods of achieving this, giving calculated values in each case.

7. A lamp emitting 450 cd in all directions is suspended 3 m above the floor. The illuminance on the floor immediately below the lamp is
   (a) 150 lx   (b) 1350 lx   (c) 50 lx   (d) 0.02 lx

8. If the lamp of question 7 is reduced in height by 0.5 m, the illuminance produced immediately below it is
   (a) 72 lx   (b) 36.7 lx   (c) 129 lx   (d) 180 lx

Mechanics

MOMENT OF FORCE

The moment of force about a point is found by multiplying together the force and the perpendicular distance between the point and the line of action of the force.

Consider an arm attached to a shaft as in Figure 85. The moment acting on the shaft tending to turn it clockwise is

$$2\,N \times 0.5\,m = 1\,Nm$$

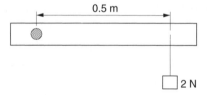

Fig. 85

TORQUE

If in Figure 85 a turning effect is applied to the shaft in the opposite direction so that the arm is maintained in a horizontal position, then the *torque* exerted at the shaft is 1 Nm.

Consider now an electric motor fitted with a pulley 0.25 m in diameter over which a belt passes to drive a machine (Figure 86). If the pull on the tight side of the belt is 60 N when the motor is running, then a continuous torque of

$$60\,N \times \frac{0.25\,m}{2} = 7.5\,Nm \text{ is present}$$

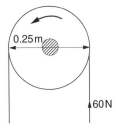

Fig. 86

This ignores any pull on the slack side of the belt, and this must usually be taken into account. Thus, if the tension in the slack side of the belt is say 10 N, then the net torque exerted by the motor is

$$(60 - 10)\,\text{N} \times \frac{0.25\,\text{m}}{2} = \frac{50 \times 0.25\,\text{Nm}}{2}$$

$$= 6.25\,\text{Nm}$$

In general the torque exerted is

$$T = (F_1 - F_2) \times r\,\text{Nm}$$

where F_1 is the tension in the tight side, F_2 is the tension in the slack side (in newtons), and r is the pulley radius (in metres).

POWER

$$P = 2\pi n T\,\text{watts}$$

where T is the torque in newton metres and n is the speed of the pulley in revolutions per second.

EXAMPLE I   If the pulley previously considered is running at 16 rev/s, calculate the power output of the motor.

$$P = 2\pi n T$$

$$= 2\pi \times 16 \times 6.25$$

$$= 629\,\text{W}$$

EXAMPLE 2 Calculate the full-load torque of a 3 kW motor running at 1200 rev/min.

$$1200 \, \text{rev/min} = \frac{1200}{60} = 20 \, \text{rev/s}$$

$$P = 2\pi n T$$

∴ $$3 \times 1000 = 2\pi \times 20 \times T$$ (note conversion of kW to W)

∴ $$T = \frac{3 \times 1000}{2\pi \times 20}$$

$$= 23.9 \, \text{Nm}$$

EXAMPLE 3 During a turning operation, a lathe tool exerts a tangential force of 700 N on the 100 mm diameter workpiece.
(a) Calculate the power involved when the work is rotating at 80 rev/min.
(b) Calculate the current taken by the 230 V single-phase a.c. motor, assuming that the lathe gear is 60% efficient, the motor is 75% efficient, and its power factor is 0.7.
The arrangement is shown in Figure 87.
(a) The torque exerted in rotating the work against the tool is

$$T = 700 \, \text{N} \times 0.05 \, \text{m}$$ (note: radius is 50 mm = 0.05 m)

$$= 35 \, \text{Nm}$$

$$P = 2\pi n T$$

$$= \frac{2\pi \times 80 \times 35}{60}$$ (note: conversion of rev/min to rev/s)

$$= 293 \, \text{W}$$

Fig. 87

(b)
$$\text{Motor output} = 293 \times \frac{100}{60}$$

$$= 488\,\text{W}$$

$$\text{Motor input} = 488 \times \frac{100}{75}$$

$$= 650.6\,\text{W}$$

$$P = V \times I \times p.f.$$

$\therefore$
$$650.6 = 230 \times I \times 0.7$$

$\therefore$
$$\text{motor current } I = \frac{650.6}{230 \times 0.7}$$

$$= 4.04\,\text{A}$$

SURFACE SPEED, PULLEY DIAMETER AND SPEED RATIOS

EXAMPLE 1   When turning a piece of low-carbon steel, it should be rotated so that the speed of its surface in relation to the tool is about 0.35 m/s. Determine the speed at which a bar 120 mm in diameter should be rotated in order to achieve this surface speed. Consider a point on the surface of the steel (Figure 88). In one revolution, this point moves through a distance equal to the circumference of the bar.

i.e.   distance moved in one revolution $= \pi \times D$

$$= 3.142 \times \frac{120}{1000}$$

$$= 0.377\,\text{m}$$

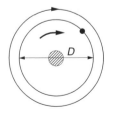

Fig. 88

$$\text{Number of revolutions required for 0.35 m} = \frac{0.35}{0.377}$$

$$= 0.9285$$

$\therefore$ $$\text{speed of rotation} = 0.928 \text{ rev/s}$$

EXAMPLE 2   A machine is driven at 6 rev/s by a belt from a standard motor running at 24 rev/s. The motor is fitted with a 200 mm diameter pulley. Find the size of the machine pulley.

The speeds at which the pulleys rotate are inversely proportional to their diameters. Thus, if the pulley having a diameter of D_1 rotates at n_1 rev/min and the pulley having a diameter of D_2 rotates at n_2 rev/min (Figure 89), then

$$\frac{n_1}{n_2} = \frac{D_2}{D_1}$$

In this case,

$$\frac{24}{6} = \frac{D_2}{200}$$

$\therefore$
$$D_2 = \frac{200 \times 24}{6}$$

$$= 800 \text{ mm}$$

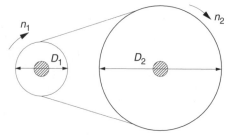

Fig. 89

EXERCISE 20

   1. A motor drives a machine by means of a belt. The tension in the tight side of the belt is 100 N, that in the slack side is

40 N, and the pulley is 200 mm in diameter. Calculate the total torque exerted by the motor.

2. A test on an induction motor fitted with a prony brake yielded the following results:

Tension in tight side of belt (N)	0	20	30	40	50	60
Tension in slack side of belt (N)	0	4	6	8.75	11.5	14.5
Speed (rev/min)	1450	1440	1430	1410	1380	1360

Calculate the torque and power corresponding to each set of readings. Take the pulley radius as being 100 mm.

3. A 10 kW motor fitted with a 250 mm diameter pulley runs at 16 rev/s. Calculate the tension in the tight side of the belt. Ignore any tension in the slack side.

4. A 4 kW motor fitted with a 150 mm diameter pulley runs at 24 rev/s. The tension in the tight side of the belt may be assumed to be equal to three times the tension in the slack side. Determine the tension in each side of the belt at full load.

5. Calculate the full-load torque of each of the motors to which the following particulars refer:

	Rated power (kW)	Normal speed (rev/min)
(a)	10	850
(b)	2	1475
(c)	18	750
(d)	0.25	1480
(e)	4	1200

6. A motor exerts a torque of 25 Nm at 16 rev/s. Assuming that it is 72% efficient, calculate the current it takes from a 440 V d.c. supply.

7. A brake test on a small d.c. motor, pulley diameter 75 mm, gave the following results:

Net brake tension (N)	0	5	10	15	20	25
Speed (rev/min)	1700	1690	1680	1670	1650	1640
Current (A)	0.8	1.05	1.3	1.68	1.9	2.25
Supply voltage (V)	116	116	116	116	116	116

For each set of values, calculate the power output and the efficiency. Plot a graph of efficiency against power.

8. The chuck of a lathe is driven at 2 rev/s through a gear which is 60% efficient from a 240 V d.c. motor. During the turning operation on a 75 mm diameter workpiece, the force on the tool is 300 N. Calculate the current taken by the motor, assuming its efficiency is 70%.

9. Calculate the speed at the circumference of a 250 mm diameter pulley when it is rotating at 11 rev/s.

10. A motor drives a machine through a vee belt. The motor pulley is 120 mm in diameter. Calculate the speed at which the belt travels when the motor runs at 24 rev/s.

11. The recommended surface speed for a certain type of grinding wheel is about 20 m/s. Determine the speed at which a 250 mm diameter wheel must rotate in order to reach this speed.

12. For a certain type of metal, a cutting speed of 0.6 m/s is found to be suitable. Calculate the most suitable speed, in revolutions per minute, at which to rotate bars of the metal having the following diameters in order to achieve this surface speed: (a) 50 mm, (b) 125 mm, (c) 150 mm, (d) 200 mm, (e) 75 mm.

13. A circular saw is to be driven at 60 rev/s. The motor is a standard one which runs at 1420 rev/min and is fitted with a 200 mm diameter pulley. Determine the most suitable size pulley for driving the saw.

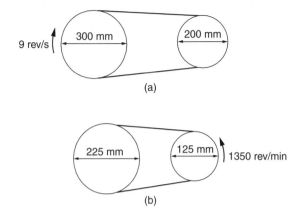

(a)

(b)

Fig. 90

14. **(a)** Calculate the speed of the smaller pulley in Figure 90(a).
    (b) Determine the speed, in rev/min, of the larger pulley in Figure 90(b).
15. Calculate the diameter of the larger pulley in Figure 91(a) and (b).

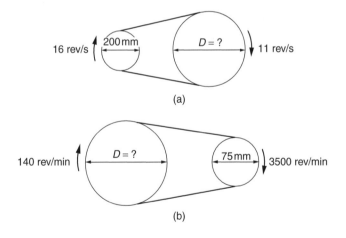

(a)

(b)

Fig. 91

16. A cutting tool exerts a tangential force of 300 N on a steel bar 100 mm in diameter which is rotating at 160 rev/min in a lathe. The efficiency of the lathe gearing is 62% and that

of the 240 V a.c. driving motor is 78%. Calculate the current taken by the motor if its power factor is 0.6.

The pulley on the lathe which takes the drive from the motor is 225 mm in diameter and rotates at 600 rev/min. The motor runs at 1420 rev/min. What is the diameter of the motor pulley?

Miscellaneous examples

$$U = E - I_a R_a$$

where U is the terminal voltage,

        E is the generated e.m.f.,

        I_a is the armature current,

and R_a is the armature resistance.

EXAMPLE   Calculate the e.m.f. generated by a shunt generator which is delivering 15 A at a terminal voltage of 440 V. The armature circuit resistance is 0.15 Ω, the resistance of the shunt field is 300 Ω, and a voltage drop of 2 V occurs at the brushes.

The circuit is shown in Figure 92.

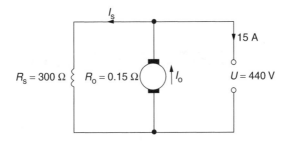

Fig. 92

To find the shunt field current,

$$U = I_s \times R_s$$

where I_s is shunt field current,

and R_s is shunt field resistance.

$$\therefore \quad 440 = I_s \times 300$$

$$\therefore \quad I_s = \frac{440}{300}$$

$$= 1.47\,\text{A}$$

Total armature current $= 15 + 1.47$

$$= 16.47\,\text{A}$$

Neglecting the voltage drop at the brushes,

$$U = E - I_a R_a$$

$$\therefore 440 = E - 16.47 \times 0.15$$

$$= E - 2.47$$

$$\therefore 440 + 2.47 = E$$

$$E = 442.47\,\text{V}$$

Allowing for the voltage drop at the brushes,

$$\text{generated e.m.f.} = 442.47 + 2$$

$$= 444.47$$

$$= 444\,\text{V}$$

D.C. MOTORS

$$U = E + I_a R_a$$

where U is the terminal voltage,

E is the back e.m.f.,

I_a is the armature current,

and R_a is the circuit resistance.

EXAMPLE Calculate the back e.m.f. of a d.c. motor which is taking an armature current of 25 A from a 220 V supply.

The resistance of its armature is $0.2\,\Omega$.

$$U = E + I_a R_a$$

$\therefore \quad 220 = E + 25 \times 0.2$

$\qquad\qquad = E + 5.0$

$\therefore \quad 220 - 5 = E$

$\therefore \qquad\qquad E = 215\,\text{V}$

ALTERNATORS AND SYNCHRONOUS MOTORS

$$f = n \times p$$

where f is the frequency in hertz,

$\qquad n$ is the speed in revolutions per second,

and p is the number of *pairs* of poles.

EXAMPLE 1   Calculate the number of poles in an alternator which generates $60\,\text{Hz}$ at a speed of $5\,\text{rev/s}$.

$\qquad f = n \times p$

$\therefore \quad 60 = 5 \times p$

$\therefore \quad p = \dfrac{60}{5}$

$\qquad\quad = 12$

$\therefore$ the machine has $2 \times p = 24$ poles

EXAMPLE 2   Calculate the speed at which a four-pole synchronous motor will run from a $50\,\text{Hz}$ supply.

$\qquad f = n \times p$

$\therefore \quad 50 = n \times 2$ (4 poles gives 2 pairs)

$\therefore \quad n = \dfrac{50}{2}$

$\qquad\quad = 25\,\text{rev/s}$

$$\text{Percentage slip} = \frac{n_s - n_r}{n_s} \times 100\%$$

where n_s is the synchronous speed,

and n_r is the actual speed of the rotor.

The synchronous speed n_s can be determined from the relationship

$$f = n_s \times p$$

as in the case of the synchronous motor.

EXAMPLE   Calculate the actual speed of a six-pole cage-induction motor operating from a 50 Hz supply with 7% slip.

$$f = n_s \times p$$

$$\therefore \quad 50 = n_s \times 3$$

$$\therefore \quad n_s = \frac{50}{3}$$

$$= 16.7 \text{ rev/s}$$

$$\text{Percentage slip} = \frac{n_s - n_r}{n_s} \times 100$$

$$\therefore \quad 7 = \frac{16.7 - n_r}{16.7} \times 100$$

$$0.07 = \frac{16.7 - n_r}{16.7}$$

$$\therefore \quad 0.07 \times 16.7 = 16.7 - n_r$$

$$\therefore \quad n_r = 16.7 - 0.07 \times 16.7$$

$$= 15.5 \text{ rev/s}$$

INSULATION RESISTANCE

The insulation resistance of a cable is *inversely* proportional to its length.

EXAMPLE 1 The insulation resistance measured between the cores of a certain twin cable 100 m long is 1000 MΩ. Calculate the insulation resistance of 35 m of the same cable.

The *shorter* length will have a *higher* value of insulation resistance because the path for the leakage current has less cross-sectional area (Figure 93).

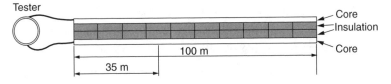

Fig. 93

Insulation resistance of 100 m = 1000 MΩ

$$\therefore \quad \text{insulation resistance of 35 m} = 1000 \times \frac{100 \, (larger)}{35 \, (smaller)}$$

$$= 2857 \, \Omega$$

EXAMPLE 2 The insulation resistance measured between the cores of a certain twin cable is 850 MΩ. Calculate the insulation resistance obtained when two such cables are connected (a) in series, (b) in parallel.

It is seen from Figure 94 that the effect in both cases is the same, i.e. to increase the c.s.a. of the leakage-current path through

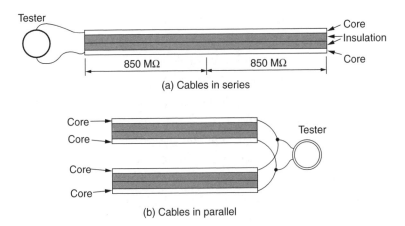

(a) Cables in series

(b) Cables in parallel

Fig. 94

the insulation. The insulation resistance in either case is thus

$$\frac{850\,\text{M}\Omega}{2} = 425\,\text{M}\Omega$$

EXERCISE 21

1. What is meant by the expression 'back e.m.f.' of a direct-current motor? In what way does the back e.m.f. affect the starting of a direct-current motor?

   A direct-current motor connected to a 460 V supply takes an armature current of 120 A on full load. If the armature circuit has a resistance of 0.25 Ω, calculate the value of the back e.m.f. at this load. (CGLI)

2. A d.c. machine has an armature resistance of 8 Ω. Calculate
   (a) the back e.m.f. when it is operating from a 110 V supply and taking an armature current of 2.5 A;
   (b) the e.m.f. generated when the machine is running as a generator and delivering 2 A at a terminal voltage of 110 V. (Neglect the field current.)

3. A d.c. motor connected to a 460 V supply has armature resistance of 0.15 Ω. Calculate

   (a) the value of the back e.m.f. when the armature current is 120 A,
   (b) the value of the armature current when the back e.m.f. is 447.4 V. (CGLI)

4. Explain briefly, with the aid of diagrams, the differences between series, shunt, and compound d.c. generators.

   A d.c. shunt generator delivers a current of 96 A at 240 V. The armature resistance is 0.15 Ω, and the field winding has a resistance of 60 Ω. Assuming a brush contact drop of 2 V, calculate (a) the current in the armature, (b) the generated e.m.f. (CGLI)

5. Calculate the speed at which an eight-pole alternator must be driven in order to generate 50 Hz.

6. Calculate the frequency of the voltage generated by a four-pole alternator when it is running at (a) 16 rev/s, (b) 12 rev/s.

7. Determine the speed at which a six-pole synchronous motor will run from the 50 Hz mains.

8. The synchronous speed of an induction motor is 750 rev/min. The motor actually runs at 715 rev/min. Calculate the percentage slip.

9. A four-pole induction motor is operating at 24 rev/s from a 50 Hz supply. Calculate the percentage slip.

10. A cage-induction motor having six poles operates with a 4.5% slip from a 50 Hz supply. Calculate the actual rotor speed.

11. Calculate the full-load torque of a 30 kW six-pole 50 Hz induction motor, assuming that the slip at full load amounts to 5%.

12. Explain the term 'insulation resistance'. Describe, with wiring diagram, a suitable instrument for measuring insulation resistance.

    Calculate the insulation resistance of a 100 m coil of insulated cable. The insulation resistance of 1 km of the same cable is given as 2500 MΩ.

13. The insulation resistance of 1000 m of two-core cable is 1500 MΩ. Calculate the insulation resistance of
    (a) 100 m **(b)** 200 m **(c)** 400 m **(d)** 600 m **(e)** 800 m
    and plot a graph showing the relationship between cable length and insulation resistance.

14. Explain the term 'insulation resistance of an installation'. Describe, with connection diagram, the working of an instrument suitable for measuring insulation resistance.

    Three separate circuits are disconnected from a distribution board and tested for insulation resistance to earth. The respective values are 40 MΩ, 60 MΩ, and 300 MΩ. What is the combined insulation resistance to earth? (CGLI)

15. The insulation resistance measured between the cores of a certain twin cable is 950 Ω. Calculate the insulation resistance of three identical cables connected in parallel.

16. The resistance of an armature circuit of a motor is 1.2 Ω. The current through it is 15 A and the terminal voltage is 200 V. The generated e.m.f. is

    (a) 218 V   (b) 182 V   (c) 13.3 V   (d) 125 V

17. An alternator generates 400 Hz at a speed of 500 rev/min. The number of pairs of poles is

    (a) 12   (b) 48   (c) 6   (d) 3

18. The insulation resistance measured between the cores of a cable is 900 MΩ for a 500 m length. The insulation resistance for 350 m of this cable would be:

    (a) 1285.7 MΩ (b) 630 MΩ (c) 157.5×10^6 MΩ (d) 194.4 MΩ

Formulae

Voltage	$U = I \times R$
Current	$I = \dfrac{U}{R}$
Resistance	$R = \dfrac{U}{I}$
Power	$P = U \times I$
Power loss	$P = I^2 R$
Current	$I = \dfrac{P}{U}$
Voltage	$U = \dfrac{P}{I}$
Resistors in parallel	$\dfrac{1}{R_1} + \dfrac{1}{R_2} + \dfrac{1}{R_3} = \dfrac{1}{R} \quad \therefore R$
Area of a circle (mm^2 or m^2)	$\dfrac{\pi \times d^2}{4} = $ c.s.a.
Circumference of a circle (mm or m)	$\pi \times d = C$
Area of triangle (mm^2 or m^2)	$\dfrac{1}{2}$base $\times$ height
Resistance of a copper conductor (Ω)	$\dfrac{1.78 \times 10^{-8} \times L}{\text{c.s.a.} \times 10^{-6}} = R$ (where c.s.a. is in mm^2)
Resistance of an aluminium conductor (Ω)	$\dfrac{2.84 \times 10^{-8} \times L}{\text{c.s.a.} \times 10^{-6}} = R$ (where c.s.a. is in mm^2)
Transformer calculation	$\dfrac{U_p}{U_s} = \dfrac{N_p}{N_s} = \dfrac{I_s}{I_p}$
Transformer efficiency	$\dfrac{\text{power out}}{\text{power in}} = $ per unit $\times$ 100 (for %)

$W = f \times d$ Work in N/m = force in Newtons × distance in mm or m 1 kg = 9.81 Newtons

$$P = \frac{W}{t} \quad \text{or} \quad \frac{\text{Work done (Nm)}}{\text{Time (secs)}} = \text{Power in watts}$$

$J = W \times t$ or   Energy (joules) = Watts × time in seconds

$$E = \frac{\text{Output}}{\text{Input}} \times 100 \quad \text{Efficiency in \%}$$

CAPACITANCE

Charge of a capacitor is in coulombs $Q = CU$

Total charge of more than one capacitor $Q = Q_1 + Q_2 + Q_3$ etc.

or capacitance is $\dfrac{Q}{U}$ Farads

Total capacitance of series connected $\dfrac{1}{C_1} + \dfrac{1}{C_2} + \dfrac{1}{C_3}$ etc. $= \dfrac{1}{C_T} = C$

Total capacitance of parallel connected $C_1 + C_2 + C_3$ etc. $= C$

Energy stored in a capacitive circuit

Energy $W = \dfrac{1}{2}CV^2$ Joules

Energy stored in an inductive circuit

Energy $W = \dfrac{1}{2}LI^2$ Joules (where L is in henrys)

THREE-PHASE CALCULATIONS

I_p = phase current

I_L = line current

U_L = line voltage

U_P = phase voltage

IN STAR (ONLY ONE CURRENT)

$$I_P = I_L$$

$$U_P = \frac{U_L}{\sqrt{3}}$$

$$U_L = U_P\sqrt{3}$$

$$P = \sqrt{3} \times U_L \times I_L$$

$$I_L = \frac{P}{\sqrt{3} \times U_L}$$

IN CIRCUITS WITH POWER FACTOR

$$P = \sqrt{3} \times U_L \times I_L \times \cos\phi$$

$$I_L = \frac{P}{\sqrt{3} \times U_L \times \cos\phi}$$

IN DELTA (ONLY ONE VOLTAGE)

$$U_L = U_P$$

$$I_P = \frac{I_L}{\sqrt{3}}$$

$$I_L = I_P \times \sqrt{3}$$

$$P = \sqrt{3} \times U_L \times I_L$$

IN CIRCUITS WITH POWER FACTOR

$$P = \sqrt{3} \times U_L \times I_L \times \cos\phi$$

$$I_L = \frac{P}{\sqrt{3} \times U_L \times \cos\phi}$$

$$\text{Power factor } \cos\theta = \frac{\text{True power}}{\text{Apparent power}} = \frac{\text{Watts}}{\text{Volt} \times \text{amps}}$$

$$Z^2 = R^2 + X^2 \quad \text{or} \quad Z = \sqrt{R^2 + X^2}$$

$$R^2 = Z^2 - X^2 \quad \text{or} \quad R = \sqrt{Z^2 - X^2}$$

$$X^2 = Z^2 - R^2 \quad \text{or} \quad X = \sqrt{Z^2 - R^2}$$

$$\text{KVA}^2 = \text{kW}^2 = \text{kVAr}^2 \quad \text{or} \quad \text{kVA} = \sqrt{\text{kW}^2 + \text{kVAr}^2}$$

$$\text{kW}^2 = \text{kVA}^2 - \text{kVAr}^2 \quad \text{or} \quad \text{kW} = \sqrt{\text{kVA}^2 - \text{kVAr}^2}$$

$$\text{kVAr}^2 = \text{kVA}^2 - \text{kW}^2 \quad \text{or} \quad \text{kVAr} = \sqrt{\text{kVA}^2 - \text{kW}^2}$$

CAPACITIVE REACTANCE

$$X_C = \frac{1}{2\pi f C \times 10^{-6}} \quad \text{or} \quad \frac{1 \times 10^6}{2\pi f C}$$

$$C = \frac{1}{2\pi f X \times 10^{-6}} \quad \text{or} \quad \frac{1 \times 10^6}{2\pi f X}$$

Inductive reactance

$$X_L = 2\pi f L$$

$$L = \frac{X_L}{2\pi f}$$

SYNCHRONOUS SPEED AND SLIP CALCULATIONS

N_S is synchronous speed in revs/sec or $\times 60$ for revs/min

N_R is speed of rotor in revs/sec or $\times 60$ for revs/min

f is frequency of supply

P is pairs of poles

Unit slip is shown as a decimal

Percentage slip is shown as %

SYNCHRONOUS SPEED

$$N_S = \frac{f}{P} \quad \text{in revs per sec} \times 60 \text{ for rpm}$$

ROTOR SPEED

$$\frac{N_S - N_R}{N_S} = \text{unit slip} \times 100 \text{ for \%}$$

CALCULATIONS ASSOCIATED WITH CABLE SELECTION

$$I_t \geq \frac{I_N}{\text{Correction factors}}$$

Cable resistance at $20\,^{\circ}C$ $\quad R = \dfrac{r_1 + r_2 \times \text{length in m}\Omega}{1000}$

Volt drop in cable $\quad \dfrac{\text{mV} \times \text{amperes} \times \text{length}}{1000}$

Earth fault loop impedance $\quad Z_s = Z_e = (R_1 + R_2)$

Glossary

a.c.	Alternating current
Area	Extent of a surface
BS 7671	British standard for electrical wiring regulations
Capacitive reactance	The effect on a current flow due to the reactance of a capacitor
Circle	Perfectly round figure
Circuit breaker	A device installed into a circuit to automatically break a circuit in the event of a fault or overload and which can be reset
Circuit	Assembly of electrical equipment which is supplied from the same origin and protected from overcurrent by a protective device
Circumference	Distance around a circle
Conductor	Material used for carrying current
Coulomb	Quantity of electrons
Correction factor	A factor used to allow for different environmental conditions of installed cables
C.S.A.	Cross-sectional area
Current	Flow of electrons
Cycle	Passage of an a.c. waveform through 360°
Cylinder	Solid or hollow, roller-shaped body
d.c.	Direct current
Dimension	Measurement
Earth fault current	The current which flows between the earth conductor and live conductors in a circuit

Earth fault loop impedance	Resistance of the conductors in which the current will flow in the event of an earth fault. This value includes the supply cable, supply transformer and the circuit cable up to the point of the fault
Efficiency	The ratio of output and input power
Energy	The ability to do work
E.M.F.	Electromotive force in volts
Frequency	Number of complete cycles per second of an alternating wave form
Fuse	A device installed in a circuit which melts to break the flow of current in a circuit
Force	Pull of gravity acting on a mass
Hertz	Measurement of frequency
Impedance	Resistance to the flow of current in an a.c. circuit
Impedance triangle	Drawing used to calculate impedance in an a.c. circuit
Internal resistance	Resistance within a cell or cells
Kilogram	unit of mass
kW	True power ($\times 1000$)
kVA	Apparent power ($\times 1000$)
kVAr	Reactive power ($\times 1000$)
Load	Object to be moved
Load	The current drawn by electrical equipment connected to an electrical circuit
Mutual induction	Effect of the magnetic field around a conductor on another conductor
Magnetic flux	Quantity of magnetism measured in Webers
Magnetic flux density	Is the density of flux measured in Webers per metre squared or Tesla
Newton	Pull of gravity (measurement of force)
On-Site Guide	Publication by the IEE containing information on electrical installation
Ohm	Unit of resistance

Overload current	An overcurrent flowing in a circuit which is electrically sound
Percentage efficiency	The ratio of input and output power multiplied by 100
Power	Energy used doing work
Pressure	Continuous force
Primary winding	Winding of transformer which is connected to a supply
Perimeter	Outer edge
Potential difference	Voltage difference between conductive parts
Prospective short circuit current	The maximum current which could flow between live conductors
Prospective fault current	The highest current which could flow in a circuit due to a fault
Protective device	A device inserted into a circuit to protect the cables from overcurrent or fault currents
Resistor	Component which resists the flow of electricity
Resistance	Opposition to the flow of current
Resistivity	Property of a material which affects its ability to conduct
Rectangle	Four-sided figure with right angles
Space factor	Amount of usable space in an enclosure
Secondary winding	Winding of transformer which is connected to a load
Self-induction	Effect of a magnetic field in a conductor
Series	Connected end to end
Thermoplastic	Cable insulation which becomes soft when heated and remains flexible when cooled down
Transpose	Change order to calculate a value
Triangle	Three-sided object
Thermosetting	Cable insulation which becomes soft when heated and is rigid when cooled down

Transformer   A device which uses electromagnetism to convert a.c. current from one voltage to another

Voltage drop   Amount of voltage lost due to a resistance

Volume   Space occupied by a mass

Wattmeter   Instrument used to measure true power

Waveform   The shape of an electrical signal

Work   Energy used moving a load (given in Newton metres or joules)

Phasor   Drawing used to calculate electrical values

Answers

1.

Volts V (a.c.)	10	225	230	400	100	25	230	625
Current (A)	0.1	15	0.5	0.4	0.01	500	180	25
Impedance (Ω)	100	15	460	1000	10000	0.05	1.3	25

2.

Current (A)	1.92	3.84	18.2	2.38	7.35	4.08	4.17	8.97
Volts V (a.c.)	4.7	7.5	225.7	230	107	228.5	400	235
Impedance (Ω)	2.45	1.95	12.4	96.3	14.56	56	96	26.2

3.

Impedance (Ω)	232	850	695.6	0.125	29.85	1050	129	4375
Volts V (a.c.)	176.3	230	400	26.5	0.194	457.8	238	245
Current (A)	0.76	0.27	0.575	212	0.0065	0.436	1.84	0.056

4. 101 Ω **5.** 1.096 A   **6.** (a) 2.18 Ω,   (b) 4.49 Ω

7. (a) 0.472 Ω, (b) 3.83 Ω, (c) 0.321 Ω, (d) 13 Ω, (e) 0.413 Ω

8. 84.3 Ω

9.

Volts V (a.c.)	61.1	105	153	193	230
Current (A)	2.3	4.2	6.12	7.35	9.2
Impedance (Ω)	26.56	25	25	26.26	25

10. (b)   **11.** (c)   **12.** (d)

1. 4.71 Ω **2.** 0.478 H

3.

Inductance (H)	0.04	0.159	0.12	0.008	0.152
Frequency (Hz)	50	50	48	90	60
Reactance (Ω)	12.57	50	36	4.5	57

4. (a) 40.8 Ω, (b) 0.13 H   **5.** (a) 15.97 A, (b) 13.07 A

6. (a) 3.77 Ω, (b) 2.2 Ω, (c) 0.141 Ω, (d) 0.11 Ω, (e) 14.1 Ω

7. (a) 0.955 H, (b) 0.0796 H, (c) 0.0462 H, (d) 0.398 H,
(e) 0.0159 H

10. 398 V   **11.** (a)   **12.** (c)

Exercise 3

1. (a) 53 Ω, (b) 127 Ω, (c) 79.6 Ω, (d) 21.2 Ω, (e) 397 Ω,
   (f) 265 Ω, (g) 12.7 Ω, (h) 33.5 Ω, (i) 199 Ω, (j) 42.4 Ω
2. (a) 13.3 μF, (b) 42.4 μF, (c) 265 μF, (d) 703 μF, (e) 88.4 μF,
   (f) 199 μF, (g) 70.8 μF, (h) 7.96 μF, (i) 106 μF, (j) 44.2 μF
3. 346 μF   4. 6.36 μF   6. 159 V
7. 207.6 μF   8. 364 V   9. 10 A
10. 15.2 A   11. (d)   12. (a)

Exercise 4

1. R 15   25   3.64   47.44   4.32   6.32   76.4   0.54
   R^2 225   625   13.25   2250   18.7   40   5837   0.735
2. X 29.8   0.68   0.16   0.95   0.4   897   233.7   0.197
   X^2 888   0.46   0.026   0.9   0.16   804609   54616   0.039
3. 6.71 A   4. 8.69 A

5. $R(\Omega)$ 14.5   140   9.63   3.5   57.6   94.8
   $X(\Omega)$ 22.8   74.6   15.68   34.7   4050   49.6
   $Z(\Omega)$ 27.02   159   18.4   34.87   4050   107

6. 232 Ω 7. 17.46 μF   8. (a) 16.9 Ω, (b) 73.3 Ω, (c) 71.3 Ω
9. 0.13 H, 115 V   10. (a) 28.75 Ω, (b) 0.122 H, (c) 47.9 Ω
11. 18.93 Ω, 15.04 Ω, 0.0479 H, 11.5 Ω 13. 69 μF
14. 0.318 H, 38.9 μF, 45.3 Hz   15. 14.57 A
16. (a) 7.47 A, (b) 127 μF   17. (c)   18. (c)

Exercise 5

1. 50 Ω 2. 40.1 Ω 3. 50 Ω 4. 198 Ω 5. 46.3 Ω
6. 231 Ω 7. 28 Ω 8. 1.09 Ω 9. 355 Ω 10. 751 Ω
11. 283 Ω 12. Approx. 500 Ω

13. Angle ϕ 30°   45°   60°   90°   52°24′   26°42′   83°12′   5°36′
    sin ϕ 0.5   0.7071   0.8660   1   0.7923   0.4493   0.9930   0.0976
    cos ϕ 0.8616   0.7071   0.5   0.0   0.6101   0.8934   0.1184   0.9952
    tan ϕ 0.5774   1   1.7321   0.0   1.2985   0.5029   8.3863   0.0981
14. Angle ϕ 33°3′   75°21′   17°15′   64°29′   27°56′   41°53′
    sin ϕ 0.5454   0.9675   0.2965   0.9025   0.4684   0.6676
    cos ϕ 0.8382   0.2529   0.9550   0.4308   0.8835   0.7445
    tan ϕ 0.6506   3.8254   0.3105   2.0949   0.5302   0.8967

15.

Angle ϕ	$21°48'$	$25°48'$	$65°30'$	$36°52'$	$36°52'$	$50°24'$	$65°20'$	$61°36'$
$\sin\phi$	0.3714	0.4352	0.91	0.6	0.6	0.7705	0.9088	0.8797
$\cos\phi$	0.9285	0.9003	0.4146	0.8	0.8	0.6374	0.4172	0.4754
$\tan\phi$	0.4000	0.4835	2.1948	0.75	0.75	1.2088	2.1778	1.8505

16.

Angle ϕ	$75°3'$	$64°16'$	$5°25'$	$38°34'$	$29°38'$	$72°24'$	$72°23'$	$71°27'$
$\sin\phi$	0.9661	0.9008	0.0946	0.6234	0.4945	0.9532	0.9531	0.9481
$\cos\phi$	0.2582	0.4341	0.9955	0.7819	0.8692	0.3020	0.3026	0.318
$\tan\phi$	3.7346	2.0752	0.0950	0.7973	0.5689	3.152	3.15	2.9814

17. $21.3\,\Omega, 20\,\Omega$ **18.** $3.95\,\Omega, 6.13\,\Omega$ **19.** $31°47'$ **20.** $90.9\,\Omega, 78\,\Omega$

21. $191\,\text{W}, 162\,\text{VAr}$ **22.** $32°51', 129\,\Omega$ **23.** $28°57'$ **24.** $66.6\,\Omega$

25. $37.6\,\Omega$ **26.** 37.6

27.

Phase angle ϕ	$75°30'$	$72°30'$	$65°6'$	$60°$	$56°37'$	$53°7'$	$45°40'$	$34°54'$
Power factor $\cos\phi$	0.25	0.3	0.421	0.5	0.55	0.6	0.699	0.82

28. (a) $17.8\,\Omega$, (b) $0.03\,\text{H}$, (c) $12.92\,\text{A}$, (d) 0.844

29. $4\,\Omega, 6.928\,\Omega, 8\,\Omega$

Exercise 6

1. $230\,\text{V}$ **2.** $31.1\,\text{A}, 14.1\,\text{A}$ **4.** $151\,\text{V}, 44°30'$ **5.** $3.63\,\text{A}$

Exercise 7

5. $248\,\text{V}$ **7.** $1029\,\Omega, 2754\,\Omega$ **8.** $7\,\text{kW}, 7.14\,\text{kVA}$ **9.** $2130\,\text{VA}$

10. $179\,\text{W}$ **11.** $5.1\,\text{A}$ **12.** $2.5\,\text{A}$ **13.** $197\,\text{V}$

Exercise 8

1. $0.47\,\text{A}\,(\text{lead})$ **2.** $3.4\,\text{A}, 27°55'\,(\text{lag}), 0.88\,(\text{lag})$,   **3.** $1.41\,\text{A}\,(\text{lag})$

4. $2.78\,\text{A}, 0.86\,(\text{lag})$ **5.** $3.49\,\text{A}, 0.92\,(\text{lag})$ **6.** $10.6\,\mu\text{F}$ **7.** $1.71\,\text{A}$

8. (b)   **9.** (c)

Exercise 9

1. $21.74\,\text{A}$; (a) $4\,\text{kW}$, (b) $3\,\text{kW}$ **2.** $131\,\text{kW}, 141\,\text{kVA}$

3. $6.72\,\text{kW}, 8.97\,\text{kVA}, 356\,\mu\text{F}$ **4.** $11.5\,\text{kVA}, 4.6\,\text{kW}, 7.09\,\text{kVAr}$

5. $13.8\,\text{kVA}, 6.9\,\text{kW}, 8.6\,\text{kVAr}, 37.39\,\text{A}$ **6.** $29.24\,\text{A}$

7. $124\,\mu\text{F}, 5.35\,\text{A}$

8. Power factor   0.7   0.75   0.8   0.85   0.9   0.95   1.0
Capacitance
required μF   1137   1624   2109   2589   3092   3619   4825

9. 31.1 A; (a) 414 μF, (b) 239 μF **10.** Approx. 15 μF **11.** (b)

12. (a) **13.** (b) **14.** (c)

Exercise 10

1. (a) 4.62 A, (b) 2561 W   **2.** (a) 1.63 A, (b) 1129 W

3. (a) $L_1 = 9.2$ A, (b) $L_2 = 17.69$ A, (c) $L_3 = 11.51$ A

4. (a) 4.6 A, (b) 3186 W   **5.** (a) 23.09 A, (b) 69.28 A

6. (a) L_1 20.99 A, (b) L_2 28.86 A, (c) L_3 24.93 A

7. (a) 7.66 A, 7.66 A, 5.3 kW, (b) 3.33 A, 23.09 A, 15.9 kW

8. (a) 19.21 A, (b) 13.3 kW

9. (a) 5.17 A, 6.2 kW, (b) 2.97 A, 2.06 kW

10. (a) 2.16 A, 0.468 Lag, 1.5 kW, (b) 6.49 A, 0.468 Lag, 4.5 kW

11. (a) 6.64 Ω, (b) 20 Ω **12.** (a) 884 μf, (b) 295 μf

13. $L_1 - L_2 = 6.66$ A, $L_2 - L_3 = 8$ A, $L_1 - L_3 = 12.6$ A, 14.1 kW

14. (a) 6.09 kW, (b) 22.6 A   **15.** (a) 7.1 kW, (b) 18.86 A

16. (a) 17.86 A, (b) 37.73 A, (c) 26.14 kW

17. (a) 17.6 kV, (b) in delta $V_L = 9.84$ A, $V_P = 5.7$ A, in star V_L and $V_P = 433.7$ A

Exercise 11

1. 5.29 A **2.** 15.35 A **3.** 17.32 A

Exercise 12

1. 385 V, 3.6%, 756 W **2.** (a) 410 V, (b) 1160 W **3.** 12.14 V

4. 95 mm^2 **5.** 467 mm^2, 500 mm^2, 4.61 V **6.** 5.8 V **7.** 70 mm^2

8. 25 mm^2 **9.** 70 mm^2 **10.** 14 A

11. (a) 17.32 A, (b) 20 A, (c) 19.2 mV/A/m, (d) 29.35 A, (e) 6 mm^2, (f) 3.79 V, (g) 25 mm

12. (a) (i) 1210 A, (ii) 0.25 s, (iii) 0.393 W, (iv) 585.2 A, (v) 3 s; (b) 0.44 Ω

13. (a) (i) 62.36 A, (ii) 63 A, (iii) 3.2 mV/A/m, (iv) 67.02 A, (v) 16 mm^2, (vi) 4.49 V, (b) (i) 0.51 Ω, (ii) 0.86 Ω (Table 41D)

14. (a) 96.51 A, (b) 100 A, (c) (i) 97.08 A, (ii) 70 mm^2, (iii) 1.82 V

15. (a) 36.08 A, (b) 40 A, (c) 42.55 A, (d) 2.92 mV/A/m,
   (e) 16 mm², (f) 8.22 V

16. (a) 17.32 A, (b) 20 A, (c) 9.6 mV/A/m, (d) 21.3 A, (e) 4 mm²,
   (f) 4.94 V, (g) 13.3 A (satisfactory), (h) Table 5C factor = 225,
   Table 5D factor = 260, satisfactory

Exercise 13

1. 40 mV   **2.** 45.2 Ω   **3.** 99975 Ω   **4.** 1.5×10^{-3} Ω

5. 9960 Ω, 149960 Ω, 249990 Ω; 40×10^{-3} Ω, 4×10^{-3} Ω   **6.** (c)

7. (d)   **8.** (a)   **9.** (b)

Exercise 14

1. (a) 44.36 A, (b) 35.3 A, (c) 36.1 A, (d) 66 A, (e) 79.3 A, (f) 8.05 A,
   (g) 20.9 A, (h) 59.1 A

2. 84%   **3.** 85.3%, 0.76   **4.** 76.9%, 0.754   **6.** 18.23 A   **7.** (d)

8. (a)

Exercise 15

1. 183.72 A. Thermal storage is probably on its own installation,
   if the shower could be on its own control, then normal 100 A
   consumer unit can be used.

2. 152.35 A. Propose that the under sink heaters be on their own
   consumer unit.

3. 230.32 A. See paragraph 2 of *IEE On-Site Guide*. This is a
   single-phase supply at present; consultation with the supplier
   would be essential. Perhaps a poly-phase supply would be
   available but could incur additional service cable costs.

4. 177.72 A (approx. 60 A per phase)

Exercise 16

1. 0.448 Ω   **2.** 0.538 Ω   **3.** 1.02 Ω   **4.** Yes

5. (a) Yes, (b) Yes   **6.** 5.52 (6 mm²)   **7.** (a) Phase 25 mm²,
   (b) C.P.C. 1.5 mm²

Exercise 17

1. 8.6 A   **2.** (a) 1.26 V, (b) 888.2 W   **3.** (a) 45 A, (b) 55.6 A, (c) 4.9 V

4. (a) 13.04 A, (b) 7.2 V, (c) 5.86 V, (d) 224.14 V

5. (a) 75.76 A, (b) 1.96 V   **6.** (a) 2.22 mV, (b) 25 mm^2, (c) 4.05 V

7. (a) 60.87 A, (b) 63 A, (c) 105.5 A, (d) 1.45 mV/A/m,
(e) 35 mm^2, (f) 5.14 V

8. (a) 43.5 A, (b) 50 A, (c) 64.73 A, (d) 0.96 mV/A/m, (e) 50 mm^2,
(f) 4.85 V

9. (a) 28.58 A, (b) 32 A, (c) 34 A, (d) 5.95 mV/A/m, (e) 16 mm^2,
(f) 4 V, (g) 32 mm conduit

10. (a) 39.13 A, (b) 4.47 mV/A/m, (c) 40 A, (d) 55.47 A,
(e) 10 mm^2, (f) 229.14 V, 227.93 V, 226.56 V

11. (a) 21.74 A, (b) 25 A, (c) 25 A, (d) 19.2 mV/A/m, (e) 4.0 mm^2,
(f) 4.18 V

12. (a) 27.95 A, (b) 9.2 V, (c) 30 A, (d) 30 A, (e) 8.22 mV/A/m,
(f) 8.16 V

13. (a) 34.78 A, (b) 40 A, (c) 40 A, (d) 6.0 mm^2, (e) 4.44 V

14. (a) 42.45 A, (b) 63.7 A, (c) 96.8 A, (d) 10 mV/A/m, (e) 50 mm^2,
(f) 3.01 V

15. (b)   **16.** (b)   **17.** (a)   **18.** (c)   **19.** (d)

Exercise 18

1. (a) 0.715 Ω, (b) 1.02 Ω   **2.** (a) 1.0 Ω, (b) 4.51 Ω

3. (a) 0.594 Ω, (b) 6.56 Ω   **4.** (a) 0.566 Ω, (b) 0.84 Ω, (c) 0.3 s

5. (c)   **6.** (c)   **7.** (b)

Exercise 19

1. (a) 15.6 lx, (b) 8 lx   **2.** 4.77 lx   **3.** 27 lx

4. 100 lx, 71.6 lx, 35.4 lx, 17.1 lx   **5.** 47.2 lx, 51.2 lx

6. (a) 56 lx, (b) new lamp of 385 cd or same lamp, new height
2.38 m

7. (c)   **8.** (a)

Exercise 20

1. 6 Nm

2.

Torque (Nm)	0	1.6	2.4	3.125	3.85	4.55
Power (W)	0	241.3	359	461	556	648

3. 795 N   **4.** 531 N, 177 N

5. (a) 112 Nm, (b) 13 Nm, (c) 229 Nm, (d) 1.61 Nm, (e) 31.8 Nm

6. 7.94 A

7.

Po (W)	0	33.2	66.0	98.4	130	161
n (%)	0	27.2	43.7	50.5	58.8	61.7

8. 1.4 A   **9.** 8.64 m/s   **10.** 9.05 m/s   **11.** 25.5 rev/s

12. (a) 229 rev/min, (b) 91.7 rev/min, (c) 76.4 rev/min,
(d) 57.3 rev/min, (e) 153 rev/min

13. 78.8 mm   **14.** (a) 13.5 rev/s, (b) 750 rev/min

15. (a) 291 mm, (b) 181 mm   **16.** 3.61 A, 95 mm

Exercise 21

1. 430 V   **2.** (a) 90 V, (b) 126 V   **3.** (a) 442 V, (b) 84 A

4. (a) 100 A, (b) 257 V   **5.** 12.5 rev/s   **6.** (a) 32 Hz, (b) 24 Hz

7. 16.7 rev/s   **8.** 4.67%   **9.** 4%   **10.** 15.9 rev/s   **11.** 301.5 Nm

12. 25 000 MΩ

13. (a) 15 000 MΩ, (b) 7500 MΩ, (c) 3750 MΩ, (d) 2500 MΩ,
(e) 1875 MΩ

14. 22.2 MΩ   **15.** 317 Ω   **16.** (b)   **17.** (b)   **18.** (b)